Nature, Properties and Management of Saline and Alkali Soils

Nature, Properties and Management of Saline and Alkali Soils

Editor

Sanjay Sawant

scitus
academics

Nature, Properties and Management of Saline and Alkali Soils

Edited by **Sanjay Sawant**

ISBN: 978-1-68117-080-0

Library of Congress Control Number: 2015935478

www.scitusacademics.com
Box No. 4766, 616 Corporate Way,
Suite 2, Valley Cottage,
NY 10989

Printed in United States of America on Acid Free Paper ∞

Contents

Preface

There is a great variety in their physical, chemical and biological properties, but they can be made to yield a good crop by adopting proper management practices and soil and water conservation measures, applying organic manures and chemical fertilizers and supplementing monsoon rain with irrigation water. But there are some extreme soil conditions which do not provide a suitable medium for the growth of plants and desirable microorganisms. Such soils are either too acidic or alkaline and special measures have to be adopted to bring them in proper condition before any crop can be raised on them satisfactory and economically. In high rainfall areas, acidic soils are formed due to leaching of bases or salts. In arid regions where rainfall is low and temperature high, soils become saline or alkaline due to alkaline due to accumulation of salts in the surface soil. Micro-organisms decompose organic matter in the soil and organic acids are continuously being formed. If these acids are not neutralized by free lime or bases to make the soil acidic It can be corrected by liming. The liming material may be ground lime stone ($CaCO3$), burnt lime (CaO) or hydrated lime Ca (OH)2- When lime is added, the soil solution becomes charged with calcium ions which replace hydrogen ions in the exchange complex. When burnt (CaO) or slaked or hydrated lime Ca (OH)2 is added to acidic soil, it changes into calcium bicarbonate, which in solution reacts with soil colloids. Example of superphosphate or ammonium sulphate which increases acidity in soil and maintains fertility of soil impoverished by leaching and cropping.

Editor

Chapter 1

Potential Use of Halophytes to Remediate Saline Soils

Mirza Hasanuzzaman[1], Kamrun Nahar[2,3], Md. Mahabub Alam[2], Prasanta C. Bhowmik[4], Md. Amzad Hossain[5], Motior M. Rahman[6], Majeti Narasimha Vara Prasad[7], Munir Ozturk[8], and Masayuki Fujita[2]

[1]Department of Agronomy, Faculty of Agriculture, Sher-e-Bangla Agricultural University, Sher-e-Bangla Nagar, Dhaka 1207, Bangladesh

[2]Laboratory of Plant Stress Responses, Department of Applied Biological Science, Faculty of Agriculture, Kagawa University, 2393 Ikenobe, Miki-cho, Kita-gun, Kagawa 761-0795, Japan

[3]Department of Agricultural Botany, Faculty of Agriculture, Sher-e-Bangla Agricultural University, Sher-e-Bangla Nagar, Dhaka 1207, Bangladesh

[4]Department of Plant, Soil, and Insect Sciences, University of Massachusetts, Amherst, MA 01003-7245, USA

[5]Subtropical Field Science Center, University of the Ryukyus, 1 Senbaru, Nishihara City, Okinawa 903-0213, Japan

[6]Institute of Biological Sciences, Faculty of Science, University of Malaya, 50603 Kuala Lumpur, Malaysia

[7]Department of Plant Sciences, University of Hyderabad, Andhra Pradesh 500 046, India

[8]Department of Biology, Ege University, Bornova, Izmir, Turkey

ABSTRACT

Salinity is one of the rising problems causing tremendous yield losses in many regions of the world especially in arid and semiarid regions. To maximize crop productivity, these areas should be brought under utilization where there are options for removing salinity or using the salt-tolerant crops. Use of salt-tolerant crops does not remove the salt and hence halophytes that have capacity to accumulate and exclude the salt can be an effective way. Methods for salt removal include agronomic practices or phytoremediation. The first is cost- and labor-intensive and needs some developmental strategies for implication; on the contrary, the phytoremediation by halophyte is more suitable as it can be executed very easily without those problems. Several halophyte species including grasses, shrubs, and trees can remove the salt from different kinds of salt-affected problematic soils through salt excluding, excreting, or accumulating by their morphological, anatomical, physiological adaptation in their organelle level and cellular level. Exploiting halophytes for reducing salinity can be good sources for meeting the basic needs of people in salt-affected areas as well. This review focuses on the special adaptive features of halophytic plants under saline condition and the possible ways to utilize these plants to remediate salinity.

INTRODUCTION

The total area of salt-affected soils in the world is 831 million hectares which includes 397 and 434 million hectares of saline and sodic soils, respectively [1]. The agricultural land is decreasing constantly due to population pressure, adverse environmental condition, continuously increasing natural calamities, and global climate change [2, 3]. More than 45 million hectares of irrigated land are affected by salt which account for 20% of total land and 1.5 million ha of land are taken out of production each year owing to high salinity levels [4, 5]; if it continues in such way, 50% of cultivable lands will be lost by the middle of the 21st century [6].

There are various reasons for salinity including natural (weathering of parent material, deposition of sea salt carried in wind and rain, inundation of coastal land by tidal water, etc.) and anthropogenic activities (rise of water table due to excessive irrigation by underground water, irrigation with salt containing water, poor drainage, etc.) [3, 7, 8]. Increased level of salinity negatively influences germination, plant growth and reproducibility, physiological processes, including photosynthesis, respiration, transpiration, membrane properties, nutrient balance, enzymatic activity, and metabolic activities, cellular homeostasis, and hormone regulation and leads to production of reactive oxygen species (ROS); and in severe stress, it leads to plant death [6, 9]. Salinity is a continuous process and its remediation is cost- and labor-intensive. It is a complex global problem that cannot be solved simply; rather a multidisciplinary approach is required. There are various ways for remediation and proper utilization of saline soils including agronomic practices, use of salt-tolerant crop varieties, and phytoremediation. Among them, phytoremediation can be cost-effective and environmentally sound technology for remediation of salt-impacted sites.

In recent years, plants having ability to remove salts from contaminated soils have been studied and identified by many researchers [10–13]. According to Flowers and Colmer [14], halophytes are characterized as plants that can survive and reproduce in environments where the salt concentration exceeds 200 mM of NaCl (~20 dS m^{-1}). These species constitute approximately 1% of the world's flora. Halophytes are plants capable of completing their life cycle under highly saline (NaCl) conditions [15]. Halophytes are also called euhalophytes because they have increased productivity with increasing salt levels and actually grow better under salinity condition than under fresh water conditions [16]. These halophytes possess special morphological and anatomical features as well as physiological processes which are well suited to cope with saline environments. The halophytes can effectively improve the saline soil as they are well adjusted in salt environment because of their diversified adaptation mechanisms including ion compartmentalization, osmotic adjustment, succulence, ion transport and uptake, antioxidant systems, maintenance of redox status, and salt inclusion or excretion [17]. There are diversified species of halophytes suited to grow in different saline regions throughout the world, namely, coastal saline soil, soils of mangrove forests, wet land, marshy land, lands of arid and semiarid regions, and agricultural fields. So, these plants can be grown in land and water containing high salt concentration, can be substitute for conventional crops, and can be good source of food, fuel, fodder, fiber, essential oils, and medicine [17]. At the same time, halophytes can be exploited as significant and major plant species bearing potential capability of desalination and restoration of saline soils and phytoremediation as well. By developing these precious strategies, unused and marginal land can be brought under cultivation and existing agricultural land will be more productive which will open a new door to sustain crop productivity. Considering the above facts, in this review, we have focused on the potential use of halophytes in remediation of salt-affected soils.

HALOPHYTES

Halophytes are defined in different ways by many scientists based on different criteria. Schimper [18] defined halophytes as the plants capable of normal growth in saline habitats and also able to thrive on "ordinary" soil. According to Stocker [19], they are plants which can tolerate salt concentrations over 0.5% at any stage of life. More simply, Dansereau [20] mentioned that plants which grow exclusively on saline soil are halophytes. Greenway and Munns [21] defined halophytes as follows: "a kind of native flora of saline soils, which contain solutions with a Psi of at least 3.3 bar, being equivalent to 70 mM monovalent salts." Plants that cannot survive in these habitats are classified as nonhalophytes. Unquestionably, this definition is not quite complete since there is a continuum from the least to the most salt-tolerant species. Some nonhalophytes can also survive in this kind of habitat and complete their life cycle, for example, sugar beet [22]. Some of the major halophytes are listed in Figure 1.

(a)

(b)

(c)

(d)

(e)

(f)

(g)

(h)

(i)

(j)

(k)

(l)

(m)

(n)

(o)

(p)

(q)

(r)

(s)

(t)

(u)

(v)

Figure 1: List of major halophytes discussed in this paper. (a) Mesembryanthemum crystallinum, (b) Suaeda australis, (c) Chenopodium album, (d) Salsola vermiculata, (e)Sarcocornia quinqueflora, (f) Portulaca oleracea, (g) Atriplex spp., (h) Allenrolfea occidentalis, (i) Tetragonia tetragonioides, (j) Salicornia europaea, (k) Sesuvium portulacastrum, (l) Crambe maritima, (m) Glycyrrhiza glabra, (n) Distichlis spicata, (o)Sporobolus virginicus, (p) Bruguiera gymnorrhiza, (q) Aegiceras corniculatum, (r)Sonneratia apetala, (s) Avicennia marina, (t) Rhizophora mucronata, (u) Plantago media, and (v) Suaeda maritima.

Based on ecological aspect, halophytes can be classified as (i) obligate, (ii) facultative, and (iii) habitat-indifferent halophytes [23]. Their growth pattern under saline condition is different (Figure 2). Obligate halophytes grow only in salty habitats. They show sufficient growth and development under high saline condition. Many plant species belonging to Chenopodiceae family fall in this category. Facultative halophytes are able to establish themselves on salty soils, but their optimum lies in a salt-free or at least low-salt condition. However, they can tolerate salt. Most Poaceae, Cyperaceae, and Brassicaceae species as well as a large number of dicotyledons like Aster tripolium, Glaux maritima, Plantago maritima, and so forth belong to this group. Plants that are indifferent toward their habitat are still able to cope with salty soils in nature. However, they usually grow on salt-free soils. They can compete with species that are sensitive towards salt and are on the other hand able to live on salty soils. Chenopodium glaucum, Myosurus minimus, and Potentilla anserina can grow in any habitat. In many species, such as Festuca rubra, Agrostis stolonifera, and Juncus bufonius, the populations living on salty soils and those on salt-free soils differ genetically [23]. However, all of these three kinds of halophytes perform better growth compared to glycophytes (Figure 2).

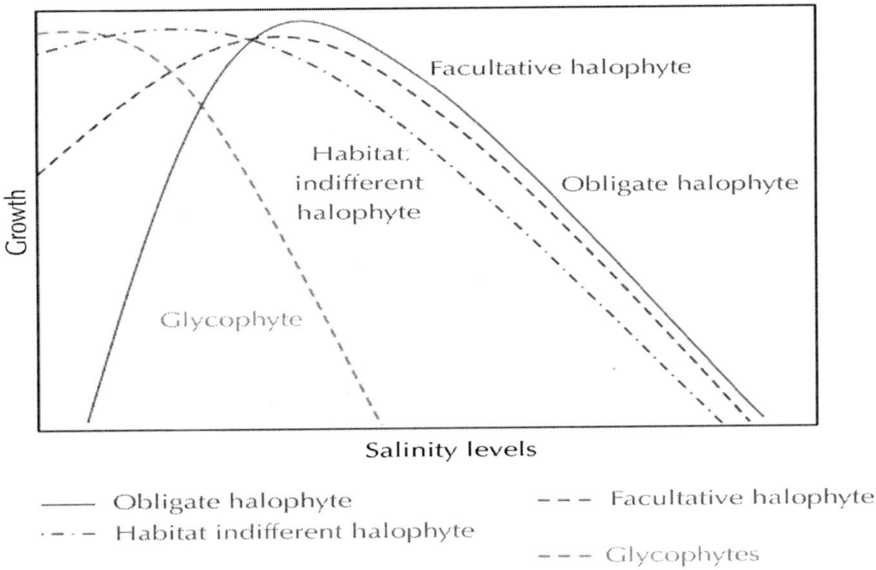

Figure 2: Possible growth pattern of halophyte under saline condition.

However, based on growth response curves, Kreeb [24] has coined four plant types (Figure 3). He stated that halophytes are those plants which exhibit maximal relative growth under some salinity in the substrate. Although the growth in halophytes is enhanced by salinity in contrast to nonhalophytes, it is still questionable whether halophytes need salt obligatorily. As obligatory halophytes would only strive successfully on saline soils the question arises if they can live on nonsaline soils [25].

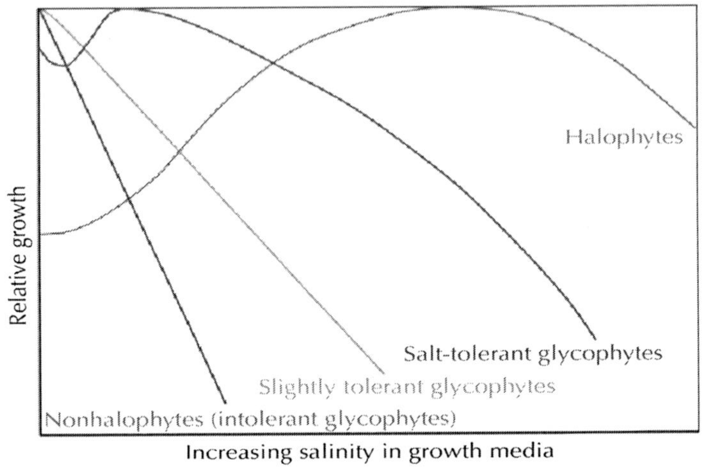

Figure 3: Schematic illustration of growth of different kinds of plants under saline condition.

MECHANISM OF ADAPTATION OF HALOPHYTES UNDER SALINE CONDITION

Salinity is not inimical to all plants. The distribution, exploitation [26], and physiology of salt tolerance of halophytes are intensively studied [16, 27, 28]. Salts taken up by halophytes do not directly control plant growth by affecting turgor, photosynthesis, or the activity of one or another enzyme. The build-up of salts in old leaves hastens their death. The loss of leaves affects the supply of assimilates or hormones to the growing organs and hereby affects growth [29, 30]. Despite their polyphyletic origins, halophytes appear to have evolved the same basic method of osmotic adjustment: accumulation of inorganic salts, mainly NaCl, in the vacuole and accumulation of organic solutes in the cytoplasm. Differences between halophyte and glycophyte ion transport systems are becoming apparent. The pathways by which Na^+ and Cl^- enter halophyte cells are not well understood but may involve ion channels and pinocytosis in addition to Na^+ and Cl^- transporters.

Na^+ uptake into vacuoles requires Na^+/H^+ antiporters in the tonoplast and H^+ ATPases and perhaps PPIases to provide the proton motive force. Tonoplast antiporters are constitutive in halophytes, whereas they must be activated by NaCl in salt-tolerant glycophytes, and they may be absent from salt-sensitive glycophytes. Halophyte vacuoles may have a modified lipid composition to prevent leakage of Na^+ back to the cytoplasm [31]. It is also to be noted that halophytes often possess large vacuoles. For example, Suaeda maritime, a potential halophyte, occupies 77% of the mesophyll cells of vacuoles [32] which makes it capable of accumulating higher concentration of salt as much as 500 mM [33]. Moreover, Na^+ concentration of the cell sap even exceeded 800 mM in another halophyte, S. maritime [34]. Although all of the halophytes exhibit better accumulation of salt, the level of total salt accumulation in the shoot is mostly species specific, depending on different adaptive strategies ([34]; Figure 4). Based on numerous studies, several adaptative mechanisms were recognized in relation to salt tolerance, which include ion compartmentalisation, osmolyte production, germination responses, osmotic adaptation, succulence, selective transport and uptake of ions, enzyme responses, salt excretion, and genetic control [35].

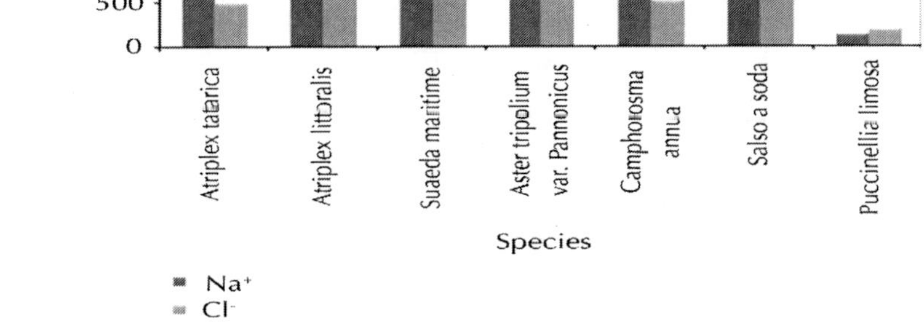

Figure 4: Na^+ and Cl^- ions concentration in the shoot of some halophytes grown in natural habitats [3] with permission from Springer.

Based on the different mechanisms of adaptation to salty condition, Walter [36] has classified the halophytes into 3 types: (i) salt excluding, (ii) salt excreting, and (iii) salt accumulating (Table 1).

Table 1: Walter's classification of halophytes [36]

Types of halophytes	Characteristics and examples
Salt excluding	In these plants, the root system possesses an ultrafiltration mechanism and this characteristic leads to establishment of such species as the dominant component of the mangrove vegetation. Example: Rhizophora mucronata, Ceriops candolleana, Bruguiera gymnorrhiza, and Kandelia candel.
Salt excreting	These plants regulate internal salt levels through foliar glands. Example: Avicennia officinalis, Avicennia alba, Avicennia marina,Aegiceros corniculatum, and Acanthus ilicifolius.
Salt accumulating	They accumulate high concentrations of salt in their cells and tissues and overcome salt toxicity by developing succulence. Example: Sonneratia apetala, Sonneratia acida, Sonneratia alba, Limnitzera racemosa, Excoecaria agallocha, Salvadora persica,Sesuvium portulacastrum, Suaeda nudiflora, and Pentatropis sianshoides.

Since ionic toxicity caused by Na^+ and Cl^- is the main concern of salt stress in plants, most studies have concentrated on Na^+ exclusion and the control of Na^+ transport within the plant [5]. Halophytes are able to tolerate high ionic concentration which involves the ability to reduce the ionic stress on the plant by minimizing the amount of Na^+ that accumulates in the cytosol of cells, particularly those in the transpiring leaves [37]. Although salt exclusion is very efficient way to minimize salt stress, the way of putting off the ions or impairing the uptake is very complex. However, true halophytes are developed with well-developed transport system that can enable a lower uptake and accumulation of salts in the upper parts of the plant, especially in the transpiring organs, especially leaves [38]. Exclusion of Na^+ happens mainly due to low net Na^+ uptake by cells in the root cortex and the tight control of net loading of the xylem by parenchyma cells in the stele [39]. Lower permeability root even under excessive concentration of soil salinity also actively helps in salt exclusion [40, 41]. There is plenty of evidence that indicated that Na^+ exclusion from leaves is associated with salt tolerance in many glycophytes including rice, wheat, and barley [42]. The capacity of salt exclusion is, however, directed by several factors like selectivity of uptake by root cells; preferential loading of K^+ rather than Na^+ into the xylem by the cells of the stele; removal of salts from the xylem in the upper parts of roots, the stem, and leaf sheaths, based upon exchange of K^+ for Na^+; and loading of the phloem [43]. The capacity of plant to sense Na^+ is also an important factor which is extracellularly done by a membrane receptor, whereas intracellular Na^+ may be sensed either by membrane proteins or by any of the many Na^+-sensitive enzymes in the cytoplasm [37].

Among several special characteristics related to the physiological adaptation of halophytes, salt excretion is one of the most efficient mechanisms that prevent excessive concentrations of salts building up in photosynthetic tissues [3]. Some of the halophytes possess multicellular salt glands and salt hairs; those are common in many halophytes such as Cressa (Convolvulaceae), Frankenia (Frankeniaceae), Spartina, Chloris, and Aeluropus (Poaceae), Atriplex (Chenopodiaceae), Statice, Limonium, Plumbago, and Armeria(Plumbaginaceae), Glaux (Primulaceae), Tamarix and Reamuria (Tamaricaceae), and some mangrove species, for example, Avicennia, Aegialitis, Aegiceras, and Acanthus [3]. These glands are composed of a set of epidermal cells complexes; those capture salt from the mesophyll cells beneath them, to which they are connected by numerous plasmodesmata, and secrete it at the leaf surface, where a layer of salt crystals is formed ([3]; Figure 5). The process of salt excretion by salt gland is yet to be elucidated by some researchers; however, one of the requisites is the availability of energy (ATP) which is required for ion pumping. In halophytes, this energy is provided by the active respiration of the glandular cells [44].

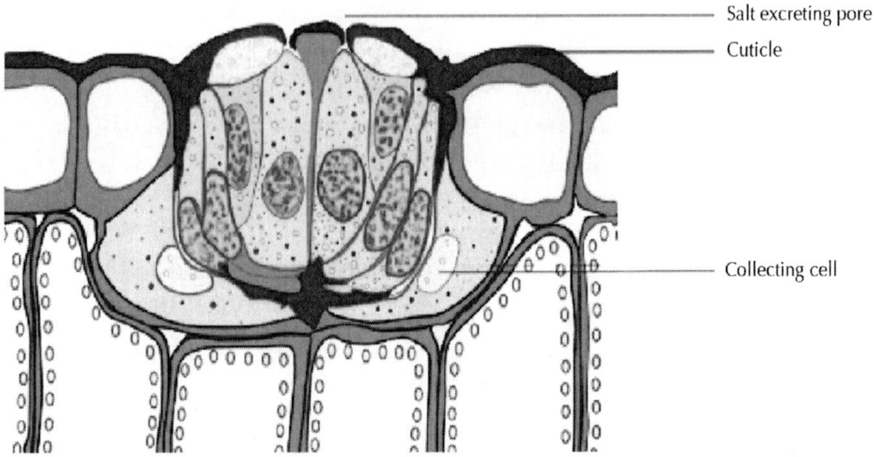

Figure 5: Cross section of a salt gland [3] with permission from Springer.

Accumulation of compatible solutes is often regarded as basic strategy for the protection and survival of halophytes under salt stress [45]. These soluble compounds, including soluble carbohydrates, GB, polyols, and Pro [46], protect plants against stress by cellular osmotic adjustment, detoxification of ROS, protection of membrane integrity, and stabilization of enzymes and proteins [47]. Moreover, the leaf tissues of halophytes are adapted to accumulate large amounts of salt ions. Such adaptive mechanism is crucial to generate a water potential gradient along root-shoot to maintain water flux throughout plants [48].

POTENTIAL USE OF HALOPHYTES UNDER SALINE CONDITION

Due to the rapid climate change, the saline area in the world is increasing day by day and currently there is an ample need to develop highly salt-tolerant crops to cope with the adverse situation. Halophytes are able to provide satisfactory yield under high salt condition. Some of halophytes producing satisfactory yield under different degrees of salinity are presented in Table 2. There are already several examples known for the utilization of halophytes for industrial, ecological, or agricultural purposes. Halophytes have been tested as vegetable, forage, and oilseed crops in agronomic field trials. The most productive species yield 10 to 20 t ha^{-1} of biomass on seawater irrigation, equivalent to conventional crops. Salicornia bigelovii, an oilseed halophyte, for example, yields 2 t ha^{-1} of seed containing 28% oil and 31% protein, which is similar to soybean yield and seed quality [31]. Many plant species have been used traditionally as herbs and vegetables and hence rediscovery of the potentials of several promising halophytic plant species to be farmed as leafy vegetables is going on for a couple of decades ([49]; Table 3). Some of the halophytes are good fodder and hence can be used for animal feeding in saline-prone areas. However, it is to be taken into consideration that some halophytes may cause nutritional barrier due to partially high salt content and antinutritional compounds [68].

Table 2: Plant species commonly found in halophytic environments and their salt tolerance limit. Adapted from Ventura and Sagi [49] with permission from Elsevier

Plant species	Salt tolerance limit	References
Aster tripolium	40 mM	Ventura and Sagi [49]
Atriplex lentiformis	500 mM	O'Leary et al. [50]
Atriplex triangularis	150 mM	Gallagher [51]
Batis maritima	500 mM	O'Leary et al. [50]
Salicornia europaea	500 mM	O'Leary et al. [50]
Salicornia persica	100 mM	Ventura et al. [52]
Sarcocornia fruticosa	100 mM	Ventura et al. [52]
Aster tripolium	300 mM	Koyro et al. [35]
Atriplex hortensis	>250 mM	Wilson et al. [53]
Batis maritima	200 mM	Debez et al. [54]
Cochlearia officinalis	100 mM	de Vos [55]
Crambe maritima	>100 mM	de Vos et al. [56]
Crithmum maritimum	150 mM	Hamed et al. [57], Ben Amor et al. [58]
Diplotaxis tenuifolia	~150 mM	de Vos [55]

Inula crithmoides	400 mM	Tardío et al. [59], Zurayk and Baalbaki [60]
Mesemyranthenum crystallinum	400 mM	Herppich et al. [61], Agarie et al. [62]
Plantago coronopus	250 mM	Koyro [63]
Portulaca oleracea	<140 mM	Simopoulos [64], Yazici et al. [65]
Salicornia sp.	>500 mM	Ventura et al. [66]
Sarcocornia sp.	>500 mM	Ventura et al. [66]
Tetragonia tetragonioides	174 mM	Wilson et al. [53], Słupski et al. [67]

Table 3: List of halophytes with highest potential as vegetable crop for saline irrigation. Adapted from Ventura and Sagi [49] with permission from Elsevier

Plant species	Salt tolerance limit	Popular uses	References
Aster tripolium	300 mM	Fresh salads, cooked vegetable	Koyro et al. [35]
Atriplex hortensis	>250 mM	Pot herb, colorful salad greens	Wilson et al. [53]
Batis maritima	200 mM	Eaten raw, cooked, or pickled	Debez et al. [54]
Cochlearia officinalis	100 mM	Fresh salads	de Vos [55]
Crambe maritima	>100 mM	Fresh salads	de Vos et al. [56]
Crithmum maritimum	150 mM	Fresh and pickled as spice and for salads	Hamed et al. [57], Ben Amor et al. [58]
Diplotaxis tenuifolia	~150 mM	Mixed salads	de Vos [55]

Inula crithmoides	400 mM	Salads, pickled in vinegar	Tardío et al. [59], Zurayk and Baalbaki [60]
Mesemyranthenum crystallinum	400 mM	Salad green or quickly cooked	Herppich et al. [61], Agarie et al. [62]
Plantago coronopus	250 mM	Salad greens	Koyro [63]
Portulaca oleracea	<140 mM	Salad greens, cooked vegetable	Simopoulos [64], Yazici et al. [65]
Salicornia sp.	>500 mM	Salad greens, vegetable	Ventura et al. [66]
Sarcocornia sp.	>500 mM	Salad greens, vegetable	Ventura et al. [66]
Tetragonia tetragonioides	174 mM	Frozen like spinach	Wilson et al. [53], Słupski et al. [67]

As the reclamation of salt-affected soils is not completely feasible and is not always cost-effective, the researchers are searching for biosaline agriculture and thus it is obvious to explore a better understanding of how naturally adapted plants (halophytes) handle salts. Study of halophytes can be useful from three perspectives [31]. First, the mechanisms by which halophytes survive and maintain productivity on saline water can be useful to develop tolerant varieties in conventional crops [69–72]. Second, halophytes grown in an agronomic setting can be used to evaluate the overall feasibility of high-salinity agriculture [31, 71]. Third, halophytes may become a direct source of new crops [12, 71, 73–76]. However, halophytes are sometimes unable to perform better in some situations. For instance, halophytes can have low biomass compared to many glycophytes [77].

Apart from accumulation of salt from the saline habitat, many of the halophytes are capable of remediating toxic metals and can grow and give yield. Halophytes are often adapted well in metal-affected habitat compared to glycophytic plants which makes them a good candidate as an ecofriendly and sustainable solution of contaminated coastal environments cleanup [78].

PHYTOREMEDIATION

Amelioration of saline and sodic soils has been predominantly achieved through the application of chemical amendments. However, amendment costs have increased prohibitively over the past two decades due to competing demands from industry and reductions in government subsidies for their agricultural use in several developing countries [79]. Since climate and cost are two vital factors in reclamation of saline land, hence, cultivation of salt-tolerant species could be an effective way to improve this situation [80]. Recently, a new environmentally safe and clean technique known as phytoremediation has been introduced to address the salinity problem. This includes the introduction of salt (ion) removing species to control salinity and to maintain the sustainability of agricultural fields [11, 12, 81]. Phytoremediation is defined as the use of plants to remove pollutants from the environment and to render them harmless [82]. These plants not only remediate the salt-contaminated soils but also provide food, fodder, fuelwood, and industrial raw material and increase the income of the farmers owning salt-affected lands. Several halophytic plant species have been tried in the past for their possible use in reclamation of salt-affected soils [81, 83–85]. After conducting number of experiments, several researchers found phytoremediation to be an effective amelioration strategy for calcareous saline-sodic and sodic soils with comparable performance against the use of chemical amendments [86–88]. Besides their positive impact on salt-affected soils, the potential use of some halophytes as forage and as oil seed crops has also been described [31]. According to Qadir et al. [79], phytoremediation has been shown to be advantageous in several aspects: (i) no financial outlay to purchase chemical amendments, (ii) accrued financial or other benefits from crops grown during amelioration, (iii) promotion of soil-aggregate stability and creation of macropores that improve soil hydraulic properties and root proliferation, (iv) greater plant-nutrient availability in soil after phytoremediation, (v) more uniform and greater zone of amelioration in terms of soil depth, and (vi) environmental considerations in terms of carbon sequestration in the postamelioration soil.

In Pakistan, Chaudhri et al. [89] investigated the ability of Suaeda fruticosa to accumulate sodium and other salts and reported that the leaves of this plant were found to contain 9.06% salt on a fresh weight basis. Ravindran et al. [81] observed that S. maritima and Sesuvium portulacastrum exhibited greater accumulation of salts in their tissue and higher reduction of salts from the saline land. It is estimated that these two halophytes could remove 504 and 474 kg of NaCl, respectively, from the saline land from 1 ha in 4-month time. Boyko [90] was the first person to suggest that halophytic plants could be used to desalinate soil and water. The hypothesis set forth by Boyko does not distinguish between sodium and other salts. Zahran and Abdel Wahid [91] made an attempt to reclaim poorly drained soils in Egypt using Juncus rigidus and J. acutusand reported that the EC of soil had a 50% saturation decreased from 33 to 22 dS m^{-1} in a single growth period. Bioreclamation of saline-sodic soil by Amshot grass (Echinochloa stagnina) in Northern Egypt, Helalia et al. [92] reported that, when compared to ponding and gypsum treatment, Amshot grass reduced the exchangeable sodium percent of the surface layer of the soil. Ke-Fu [93] found that Suaeda salsa produces about 20 tons dry weight ha^{-1} containing 3.4 tons of salt. Hamidov et al. [94] reported that Portulaca oleraceaaccumulated highest salt uptake (497 kg ha^{-1}) with biomass production of 3948 kg ha^{-1}. Rabhi et al. [11] reported that Arthrocnemum indicum, Suaeda fruticosa, and Sesuvium portulacastrum seedlings grown on a saline soil significantly reduced the soil salinity and EC by absorbing soluble salts mainly sodium ions and they also reported that Sesuvium portulacastrum was able to accumulate nearly 30% of Na^+ content in shoot over the 170 d. Nasir [95] conducted a field study in Jordan valley to investigate the effects of growing three types of salt accumulator halophyte species, Tamarix aphylla, Atriplex nummularia, and A. halimus, on chemical properties of saline sodic soil and these halophytic species decreased the soil salinity at the end of the experiment. Rabhi et al. [96] observed that Sesuvium portulacastrum, an obligate halophyte, decreased the soil salinity and sodicity.

Vegetative bioremediation or bioreclamation of salt-affected soils is an economic solution mainly for developing countries since chemical additions are becoming increasingly expensive. Several authors [13, 81,91–93, 97–100] have proved that the potential of halophytic plants to accumulate enormous salt quantities depends often on the capacity of their above ground biomass (hyperaccumulating plants). This ability could be of great importance, particularly in arid and semiarid regions, where insufficient precipitations and inappropriate systems [101] are unable to reduce the salt burden in the rhizosphere of plants [102] and suitable physicochemical methods are too expensive. Environmentally safe and clean technique to address the salinity problem includes the introduction of salt (ion) removing species to control salinity and to maintain the sustainability of agricultural fields. Large-scale decontamination of soils and underground water using phytoremediation techniques requires plants with high salt uptake rates, large biomass, and tolerance to a wide array of environmental conditions and constraints. Furthermore, salt marshes, especially salt accumulating halophytes, are dominant crop in the coastal region and introduction of these salt removing halophytic species could potentially create both environmental and economic solutions to remediate saline soils. After reclamation studies are over, these cultivated halophytes can be utilized as animal fodder or for making organic composts.

Akhter et al. [80] evaluated the phytoremediation performance of salt-tolerant species Leptochloa fusca (L.) Kunth (kallar grass) in salt-affected soils. They observed that soil salinity, sodicity, and pH decreased exponentially by growing kallar grass as a result of leaching of salts from surface (0–20 cm) to lower depths (>100 cm). Concentrations of soluble cations (Na^+, K^+, Ca^{2+}, and Mg^{2+}) and anions (Cl^-,SO_4^{2-}, and HCO_3^-) were reduced through to greater soil depths. The decline in soil pH was attributed to release of CO_2 by grass roots and solubilization of $CaCO_3$. The ameliorative effects on the soil chemical environment were pronounced after three years of growing kallar grass. Cultivation of kallar grass enhanced leaching and interactions among soil chemical properties and thus restored soil fertility. The soil maintained the improved characteristics with further growth of the grass up to five years suggesting that growing salt-tolerant plants is a sustainable approach to biological amelioration of saline wastelands [80]. Very recently, de Souza et al. [103] reported Atriplex nummularia Lindl. as a very potential halophyte that sustains under water-stressed condition on sodic or saline soil. They concluded that the growth patterns and anatomical changes shown by the halophyte A. nummularia Lindl. grown under different soil moisture conditions can contribute significantly to the management of soil and water in semiarid regions.

Phytoremediation could become a cost-effective and environmentally sound technology for remediation of salt-impacted sites if it can be properly developed. There are certain limitations that must be overcome for this plant-based remediation system to come into common usage. Phytoremediation can be time-consuming because it requires several growing seasons to lower the level of contaminants in soil. It is also limited to soil depths that are in the rooting zone [104]. Furthermore, successful remediation of soil with high levels of salt is hard to achieve by the fact that plant growth and germination are inhibited by salinity. As a result, finding salt-tolerant plants that have deep and vigorous root growth, as well as sufficient above-ground biomass production, is one of the basic criteria for the selection of plants for remediation of salt-impacted sites. However, selecting suitable plant species for the phytoremediation of salt-affected soils is also important. According to Qadir and Oster [105], plants having capability to remove the maximum quantity of salts by producing higher biomass with some economic importance are mainly selected for phytoremediation [105]. The selected plant species should tolerate high salt concentration. The plants which provide food and fodders are very effective to be used for phytoremediation. Several plant species including grasses, shrubs, and trees are being used for phytoremediation of salinized soils. Many trees have also been recommended. Phytoremediation with trees and grasses is beneficial because these can be utilized as fodder, timber, and fuel [12, 106, 107].

However, use of halophytes for soil reclamation is still in an exploratory stage and only a few field studies for bioreclamation of saline soil using halophytes have been carried out so far and therefore, more research is needed to study the utilization of halophytes to remove excess salinity added by irrigation.

CONCLUSIONS AND FUTURE PERSPECTIVES

It is clear that salinity problem is increasing rapidly throughout the world. More than half a billion hectares of land are not being properly used for crop production. Thus, there is a need to search for means to improve saline soils so that such soils could support highly productive and meaningful land-use systems to meet the current challenges of global food security. In addition, the crop adaptability to saline conditions should also be improved. Despite the importance of salinity in shaping the composition of coastal plant communities, our knowledge about how different species respond physiologically to variable salinities is limited. In particular, our understanding of physiological/biochemical mechanisms underlying halophytes under variable salinities is very scarce. Hence, the physiological and molecular studies to reveal the underlying mechanisms of these processes are important. In addition, discovering the induction of signaling cascades leading to profound changes in specific gene expression is also considered an important salt stress adaptation. Molecular knowledge of response and tolerance mechanisms will pave the way for engineered plants that can tolerate salt stress and could be the basis for production of crops which can result in economic yield under salt-stress conditions. In recent years, phytoremediation of saline soils has been studied by researchers and it was observed that the use of some halophytes could remove salt from soil. Phytoremediation could become a cost-effective and environmentally sound technology for remediation of salt-impacted sites if it is properly developed. There are certain limitations that must be overcome for this plant-based remediation system to come into common usage. Phytoremediation can be time-consuming because it requires several growing seasons to lower the level of contaminants in soil. It is also limited to soil depths that are in the rooting zone. It is necessary to find the plants having capability to remove the maximum quantity of salts by producing higher biomass with some economic importance are mainly selected for phytoremediation and the selected plant species should tolerate high salt concentration. The forthcoming challenge for using halophytes to remediate soil salinity is to develop a plant with diverse salt accumulating capacity in a cost-effective way. Identification of novel genes with high biomass yield characteristics and the subsequent development of transgenic plants with superior remediation features would be crucial for such type of research.

ACKNOWLEDGMENTS

The authors wish to thank Mr. Md. Hasanuzzaman, Department of Agronomy, Sher-e-Bangla Agricultural University, Dhaka, Bangladesh, for his critical reading of the paper draft.

REFERENCES

1. FAO, "Global Network on Integrated Soil Management for Sustain-Able Use of Salt-Affected Soils," Rome, Italy, 2000, http://www.fao.org/ag/agl/agll/spush.
2. M. Hasanuzzaman, K. Nahar, and M. Fujita, "Plant response to salt stress and role of exogenous protectants to mitigate salt-induced damages," in Ecophysiology and Responses of Plants Under Salt Stress, P. Ahmad, M. M. Azooz, and M. N. V. Prasad, Eds., pp. 25–87, Springer, New York, NY, USA, 2013.
3. M. Hasanuzzaman, K. Nahar, M. Fujita et al., "Enhancing plant productivity under salt stress—relevance of poly-omics," in Salt Stress in Plants: Omics, Signaling and Responses, P. Ahmad, M. M. Azooz, and M. N. V. Prasad, Eds., pp. 113–156, Springer, Berlin, Germany, 2013.
4. M. G. Pitman and A. Läuchli, "Global impact of salinity and agricultural ecosystem," in Salinity: Environment—Plants—Molecules, A. Läuchli and U. Lüttge, Eds., pp. 3–20, Kluwer Academic, Dodrecht, The Netherlands, 2002.
5. R. Munns and M. Tester, "Mechanisms of salinity tolerance," Annual Review of Plant Biology, vol. 59, pp. 651–681, 2008.
6. S. Mahajan and N. Tuteja, "Cold, salinity and drought stresses: an overview," Archives of Biochemistry and Biophysics, vol. 444, no. 2, pp. 139–158, 2005.
7. R. Munns, "Genes and salt tolerance: bringing them together," New Phytologist, vol. 167, no. 3, pp. 645–663, 2005.
8. G. Manchanda and N. Garg, "Salinity and its effects on the functional biology of legumes," Acta Physiologiae Plantarum, vol. 30, no. 5, pp. 595–618, 2008.
9. M. Hasanuzzaman, M. A. Hossain, J. A. Teixeira da Silva, and M. Fujita, "Plant responses and tolerance to abiotic oxidative stress: antioxidant defense is a key factor," in Crop Stress and Its Management: Perspectives and Strategies, V. Bandi, A. K. Shanker, C. Shanker, and M. Mandapaka, Eds., pp. 261–316, Springer, Berlin, Germany, 2012.
10. M. Y. Ashraf, M. Ashraf, and G. Sarwar, "Physiological approaches to improving plant salt tolerance," in Crops: Growth, Quality

and Biotechnology, R. Dris, Ed., pp. 1206–1227, WFL Publisher, Helsinki, Finland, 2005.

11. M. Rabhi, O. Talbi, A. Atia, A. Chedly, and A. Smaoui, "Selection of halophyte that could be used in the bio reclamation of salt affected soils in arid and semi-arid regions," in Biosaline Agriculture and High Salinity Tolerance, pp. 242–246, 2008.
12. M. Y. Ashraf, M. Ashraf, K. Mahmood, J. Akhter, F. Hussain, and M. Arshad, "Phytoremediation of saline soils for sustainable agricultural productivity," in Plant Adaptation and Phytoremediation, M. Ashraf, M. Ozturk, and M. S. A. Ahmad, Eds., pp. 335–3355, Springer, Berlin, Germany, 2010.
13. M. Rabhi, C. Hafsi, A. Lakhdar et al., "Evaluation of the capacity of three halophytes to desalinize their rhizosphere as grown on saline soils under nonleaching conditions," African Journal of Ecology, vol. 47, no. 4, pp. 463–468, 2009.
14. T. J. Flowers and T. D. Colmer, "Salinity tolerance in halophytes," New Phytologist, vol. 179, no. 4, pp. 945–963, 2008.
15. J. R. Stuart, M. Tester, R. A. Gaxiola, and T. J. Flowers, "Plants of saline environments," in Access Science, 2012, http://www.accessscience.com.
16. N. P. Yensen, "Halophyte uses for the twenty-first century," in Ecophysiology of High Salinity Tolerant Plants, M. A. Khan and D. J. Weber, Eds., pp. 367–396, 2008.
17. V. H. Lokhande and P. Suprasanna, "Prospects of halophytes in understanding and managing abiotic stress tolerance," in Environmental Adaptations and Stress Tolerance of Plants in the Era of Climate Change, P. Ahmad and M. N. V. Prasad, Eds., pp. 29–56, Springer, New York, NY, USA, 2012.
18. A. F. W. Schimper, Plant Geography upon a Physiological Basis, Clarendon Press, Oxford, UK, 1903.
19. O. Stocker, "Das Halophytenproblem," in Ergebnisse der Biologie, K. V. Frisch, R. Goldschmidt, W. Ruhland, and H. Winterstein, Eds., pp. 266–353, Springer, Berlin, Germany, 1928, (German).
20. P. Dansereau, Biogeography: An Ecological Perspective, Ronald Press, New York, NY, USA, 1957.
21. H. Greenway and R. Munns, "Mechanisms of salt tolerance in non halophytes," Annual Review of Plant Physiology, vol. 31, pp. 149–190, 1980.

22. Z. Kefu, F. Hai, and I. A. Ungar, "Survey of halophyte species in China," Plant Science, vol. 163, no. 3, pp. 491–498, 2002.
23. P. von Sengbusch, "Halophytes" Botanik Online, University of Hamburg, 2003.
24. K. Kreeb, "Plants in saline habitats," Naturwissenschaften, vol. 61, no. 8, pp. 337–343, 1974.
25. S. W. Breckle, "Salinity, halophytes and salt affected natural ecosystems," in Salinity: Environment—Plants—Molecules, A. Läuchli and U. Lüttge, Eds., pp. 53–77, Kluwer Academic, Dodrecht, The Netherlands, 2002.
26. Z. Kefu, F. Hai, and I. A. Ungar, "Survey of halophyte species in China," Plant Science, vol. 163, no. 3, pp. 491–498, 2002.
27. H. Koca, M. Bor, F. Özdemir, and I. Türkan, "The effect of salt stress on lipid peroxidation, antioxidative enzymes and proline content of sesame cultivars," Environmental and Experimental Botany, vol. 60, no. 3, pp. 344–351, 2007.
28. E. C. da Silva, R. J. M. C. Nogueira, F. P. de Araújo, N. F. de Melo, and A. D. de Azevedo Neto, "Physiological responses to salt stress in young umbu plants," Environmental and Experimental Botany, vol. 63, no. 1–3, pp. 147–157, 2008.
29. R. Munns, "Physiological processes limiting plant growth in saline soils: some dogmas and hypotheses," Plant, Cell & Environment, vol. 16, no. 1, pp. 15–24, 1993.
30. R. Munns, D. P. Schachtman, and A. G. Condon, "The significance of a two-phase growth response to salinity in wheat and barley," Australian Journal of Plant Physiology, vol. 22, no. 4, pp. 561–569, 1995.
31. E. P. Glenn, J. J. Brown, and E. Blumwald, "Salt tolerance and crop potential of halophytes," Critical Reviews in Plant Sciences, vol. 18, no. 2, pp. 227–255, 1999.
32. M. A. Hajibagheri, J. L. Hall, and T. J. Flowers, "Stereological analysis of leaf cells of the halophyteSuaeda maritima (L.) dum," Journal of Experimental Botany, vol. 35, no. 10, pp. 1547–1557, 1984.
33. M. N. H. Dracup and H. Greenway, "A procedure for isolating vacuoles from leaves of the halophyteSuaeda maritime," Plant, Cell & Environment, vol. 8, pp. 149–154, 1985.

34. Z. Dajic, studija halofitske zajednice Puccinellietum limosae (Rapcs.) Wend. (Ecological study of halophytic community Puccinellietum limosae (Rapcs.) Wend.) [Doctoral dissertation], Faculty of Biology, University of Belgrade, 1996.

35. H.-W. Koyro, M. A. Khan, and H. Lieth, "Halophytic crops: a resource for the future to reduce the water crisis?" Emirates Journal of Food and Agriculture, vol. 23, no. 1, pp. 1–16, 2011.

36. H. Walter, "Salinity problems in the acid zones. The adaptations of plants to saline soils," Arid Zone Research, vol. 14, pp. 65–68, 1961.

37. P. Carillo, M. Grazia Annunziata, G. Pontecorvo, A. Fuggi, and P. Woodrow, "Salinity stress and salt tolerance," in Abiotic Stress in Plants—Mechanisms and Adaptations, A. K. Shanker and B. Venkateswarlu, Eds., pp. 21–38, InTech, Rijeka, Croatia, 2011.

38. Z. Dajic, "Salt stress," in Physiology and Molecular Biology of Stress Tolerance in Plant, K. V. Madhava Rao, A. S. Raghavendra, and K. Janardhan Reddy, Eds., pp. 41–99, Springer, Amsterdam, The Netherlands, 2006.

39. R. Davenport, R. A. James, A. Zakrisson-Plogander, M. Tester, and R. Munns, "Control of sodium transport in durum wheat," Plant Physiology, vol. 137, no. 3, pp. 807–818, 2005.

40. T. J. Flowers and M. A. Hajibagheri, "Salinity tolerance in Hordeum vulgare: ion concentrations in root cells of cultivars differing in salt tolerance," Plant and Soil, vol. 231, no. 1, pp. 1–9, 2001.

41. J.-K. Zhu, "Plant salt tolerance," Trends in Plant Science, vol. 6, no. 2, pp. 66–71, 2001.

42. R. A. James, C. Blake, C. S. Byrt, and R. Munns, "Major genes for Na^+ exclusion, Nax1 and Nax2(wheat HKT1;4 and HKT1;5), decrease Na^+ accumulation in bread wheat leaves under saline and waterlogged conditions," Journal of Experimental Botany, vol. 62, no. 8, pp. 2939–2947, 2011.

43. R. Munns, "Comparative physiology of salt and water stress," Plant, Cell and Environment, vol. 25, no. 2, pp. 239–250, 2002.

44. K. B. Marcum and C. L. Murdoch, "Salt tolerance of the coastal salt marsh grass, Sporobulus virginicus(L.) Kunth," New Phytologist, vol. 120, pp. 281–288, 1992.

45. G. Lee, R. N. Carrow, R. R. Duncan, M. A. Eiteman, and M. W. Rieger, "Synthesis of organic osmolytes and salt tolerance mechanisms in Paspalum vaginatum," Environmental and Experimental Botany, vol. 63, no. 1–3, pp. 19–27, 2008.
46. A. K. Parida and A. B. Das, "Salt tolerance and salinity effects on plants: a review," Ecotoxicology and Environmental Safety, vol. 60, no. 3, pp. 324–349, 2005.
47. M. Ashraf and M. R. Foolad, "Roles of glycine betaine and proline in improving plant abiotic stress resistance," Environmental and Experimental Botany, vol. 59, no. 2, pp. 206–216, 2007.
48. J. A. G. Silveira, S. A. M. Araújo, J. P. M. S. Lima, and R. A. Viégas, "Roots and leaves display contrasting osmotic adjustment mechanisms in response to NaCl-salinity in Atriplex nummularia,"Environmental and Experimental Botany, vol. 66, no. 1, pp. 1–8, 2009.
49. Y. Ventura and M. Sagi, "Halophyte crop cultivation: the case for salicornia and sarcocornia,"Environmental and Experimental Botany, vol. 92, pp. 144–153, 2013.
50. J. W. O'Leary, E. P. Glenn, and M. C. Watson, "Agricultural production of halophytes irrigated with seawater," Plant and Soil, vol. 89, no. 1–3, pp. 311–321, 1985.
51. J. L. Gallagher, "Halophytic crops for cultivation at seawater salinity," Plant and Soil, vol. 89, no. 1–3, pp. 323–336, 1985.
52. Y. Ventura, W. A. Wuddineh, M. Shpigel et al., "Effects of day length on flowering and yield production of Salicornia and Sarcocornia species," Scientia Horticulturae, vol. 130, no. 3, pp. 510–516, 2011.
53. C. Wilson, S. M. Lesch, and C. M. Grieve, "Growth stage modulates salinity tolerance of New Zealand spinach (Tetragonia tetragonioides Pall.) and red orach (Atriplex hortensis L.)," Annals of Botany, vol. 85, no. 4, pp. 501–509, 2000.
54. A. Debez, D. Saadaoui, I. Slama, B. Huchzermeyer, and C. Abdelly, "Responses of Batis maritimaplants challenged with up to two-fold seawater NaCl salinity," Journal of Plant Nutrition and Soil Science, vol. 173, no. 2, pp. 291–299, 2010.
55. A. C. de Vos, Sustainable exploitation of saline resources: ecology, ecophysiology and cultivation of potential halophyte

crops [Ph.D. dissertation], Vrije Universiteit, Amsterdam, The Netherlands, 2011.

56. A. C. de Vos, R. Broekman, M. P. Groot, and J. Rozema, "Ecophysiological response of Crambe maritima to airborne and soil-borne salinity," Annals of Botany, vol. 105, no. 6, pp. 925–937, 2010.

57. K. B. Hamed, A. Debez, F. Chibani, and C. Abdelly, "Salt response of Crithmum maritimum, an oleagineous halophyte," Tropical Ecology, vol. 45, no. 1, pp. 151–159, 2004.

58. N. Ben Amor, K. Ben Hamed, A. Debez, C. Grignon, and C. Abdelly, "Physiological and antioxidant responses of the perennial halophyte Crithmum maritimum to salinity," Plant Science, vol. 168, no. 4, pp. 889–899, 2005.

59. J. Tardío, M. Pardo-De-Santayana, and R. Morales, "Ethnobotanical review of wild edible plants in Spain," Botanical Journal of the Linnean Society, vol. 152, no. 1, pp. 27–71, 2006.

60. R. A. Zurayk and R. Baalbaki, "Inula crithmoides: a candidate plant for saline agriculture," Arid Soil Research and Rehabilitation, vol. 10, no. 3, pp. 213–223, 1996.

61. W. B. Herppich, S. Huyskens-Keil, and M. Schreiner, "Effects of saline irrigation on growth, physiology and quality of Mesembryanthemum crystallinum L., a rare vegetable crop," Journal of Applied Botany and Food Quality, vol. 82, no. 1, pp. 47–54, 2008.

62. S. Agarie, T. Shimoda, Y. Shimizu et al., "Salt tolerance, salt accumulation, and ionic homeostasis in an epidermal bladder-cell-less mutant of the common ice plant Mesembryanthemum crystallinum," Journal of Experimental Botany, vol. 58, no. 8, pp. 1957–1967, 2007.

63. H.-W. Koyro, "Effect of salinity on growth, photosynthesis, water relations and solute composition of the potential cash crop halophyte Plantago coronopus (L.)," Environmental and Experimental Botany, vol. 56, no. 2, pp. 136–146, 2006.

64. A. P. Simopoulos, "Omega-3 fatty acids and antioxidants in edible wild plants," Biological Research, vol. 37, no. 2, pp. 263–277, 2004.

65. I. Yazici, I. Türkan, A. H. Sekmen, and T. Demiral, "Salinity tolerance of purslane (Portulaca oleraceaL.) is achieved by

enhanced antioxidative system, lower level of lipid peroxidation and proline accumulation," Environmental and Experimental Botany, vol. 61, no. 1, pp. 49–57, 2007.

66. Y. Ventura, W. A. Wuddineh, M. Myrzabayeva et al., "Effect of seawater concentration on the productivity and nutritional value of annual Salicornia and perennial Sarcocornia halophytes as leafy vegetable crops," Scientia Horticulturae, vol. 128, no. 3, pp. 189–196, 2011.

67. J. Słupski, J. Achrem-Achremowicz, Z. Lisiewska, and A. Korus, "Effect of processing on the amino acid content of New Zealand spinach (Tetragonia tetragonioides Pall. Kuntze)," International Journal of Food Science and Technology, vol. 45, no. 8, pp. 1682–1688, 2010.

68. M. A. Khan, R. Ansari, H. Ali, B. Gul, and B. L. Nielsen, "Panicum turgidum, a potentially sustainable cattle feed alternative to maize for saline areas," Agriculture, Ecosystems and Environment, vol. 129, no. 4, pp. 542–546, 2009.

69. T. Rausch, M. Kirsch, R. Low, A. Lehr, R. Viereck, and A. N. Zhigang, "Salt stress responses of higher plants: the role of proton pumps and Na^+/H^+-antiporters," Journal of Plant Physiology, vol. 148, no. 3-4, pp. 425–433, 1996.

70. R. Serrano, "Salt tolerance in plants and microorganisms: toxicity targets and defense responses,"International Review of Cytology, vol. 165, pp. 1–52, 1996.

71. E. Glenn, S. Miyamoto, D. Moore, J. J. Brown, T. L. Thompson, and P. Brown, "Water requirements for cultivating Salicornia bigelovii Torr. with seawater on sand in a coastal desert environment," Journal of Arid Environments, vol. 36, no. 4, pp. 711–730, 1997.

72. J.-K. Zhu, P. M. Hasegawa, and R. A. Bressan, "Molecular aspects of osmotic stress in plants," Critical Reviews in Plant Sciences, vol. 16, no. 3, pp. 253–277, 1997.

73. F. A. Llerenav, "Massive propagation of halophytes (Distichlis spicata and Tamarix spp.) on the highly saline-alkaline soils in the ex-Lake Texcoco, Mexico," in Halophytes As a Resource for Livestock and for Rehabilitation of Degraded Lands: Proceedings of the International Workshop on Halophytes for Reclamation of Saline Wastelands and As a Resource for Livestock—Problems and Prospects; 1992 November 22-27; Nairobi, Kenya, V. R.

Squires and A. T. Ayoub, Eds., pp. 289–292, Kluwer Academic, Boston, Mass, USA, 1994.

74. J. O'Leary, "The agricultural use of native plants on problem soils," Monographs on Theoretical and Applied Genetics, vol. 21, pp. 127–143, 1994.
75. V. R. Squires and A. T. Ayoub, Halophytes As a Resource for Livestock and for Rehabilitation of Degraded Lands, Kluwer Academic, Dodrecht, The Netherlands, 1994.
76. R. Choukr-Allah, C. V. Malcolm, and A. Hamdy, Halophytes and Biosaline Agriculture, 1996.
77. H. Zaier, T. Ghnaya, A. Lakhdar et al., "Comparative study of Pb-phytoextraction potential in Sesuvium portulacastrum and Brassica juncea: tolerance and accumulation," Journal of Hazardous Materials, vol. 183, no. 1–3, pp. 609–615, 2010.
78. N. A. Anjum, I. Ahmad, M. Válega et al., "Salt marsh halophyte services to metal-metalloid remediation: assessment of the processes and underlying mechanisms," Critical Reviews in Environmental Science and Technology, 2014.
79. M. Qadir, J. D. Oster, S. Schubert, A. D. Noble, and K. L. Sahrawat, "Phytoremediation of sodic and saline-sodic soils," Advances in Agronomy, vol. 96, pp. 197–247, 2007.
80. J. Akhter, K. Mahmood, K. A. Malik, S. Ahmed, and R. Murray, "Amelioration of a saline sodic soil through cultivation of a salt-tolerant grass Leptochloa fusca," Environmental Conservation, vol. 30, no. 2, pp. 168–174, 2003.
81. K. C. Ravindran, K. Venkatesan, V. Balakrishnan, K. P. Chellappan, and T. Balasubramanian, "Restoration of saline land by halophytes for Indian soils," Soil Biology and Biochemistry, vol. 39, no. 10, pp. 2661–2664, 2007.
82. D. E. Salt, R. D. Smith, and I. Raskin, "Phytoremediation," Annual Review of Plant Biology, vol. 49, pp. 643–668, 1998.
83. A. J. de Villiers, M. W. van Rooyen, G. K. Theron, and A. S. Claassens, "Removal of sodium and chloride from a saline soil by Mesembryanthemum barklyi," Journal of Arid Environments, vol. 29, no. 3, pp. 325–330, 1995.
84. B. Gul, D. J. Weber, and M. A. Khan, "Effect of salinity and planting density on physiological responses of Allenrolfea occidentalis,"

Western North American Naturalist, vol. 60, no. 2, pp. 188–197, 2000.

85. M. N. Jithesh, S. R. Prashanth, K. R. Sivaprakash, and A. K. Parida, "Antioxidative response mechanisms in halophytes: their role in stress defence," Journal of Genetics, vol. 85, no. 3, pp. 237–254, 2006.

86. K. Singh, H. S. Chauhan, D. K. Rajput, and D. V. Singh, "Report of a 60 month study on litter production, changes in soil chemical properties and productivity under Poplar (P. deltoides) and Eucalyptus (E. hybrid) interplanted with aromatic grasses," Agroforestry Systems, vol. 9, no. 1, pp. 37–45, 1989.

87. N. Ahmad, R. H. Qureshi, and M. Qadir, "Amelioration of a calcareous saline-sodic soil by gypsum and forage plants," Land Degradation & Rehabilitation, vol. 2, no. 4, pp. 277–284, 1990.

88. M. Qadir, R. H. Qureshi, and N. Ahmad, "Reclamation of a saline-sodic soil by gypsum and Leptochloa fusca," Geoderma, vol. 74, no. 3-4, pp. 207–217, 1996.

89 I. I. Chaudhri, B. H. Shah, N. Naqvi, and I. A. Mallick, "Investigations on the role of Suaeda fruticosaForsk in the reclamation of saline and alkaline soils in West Pakistan plains," Plant and Soil, vol. 21, no. 1, pp. 1–7, 1964.

90. H. Boyko, "Basic ecological principles of plant growing by irrigation with highly saline or seawater," inSalinity and Aridity, H. Boyko, Ed., Dr. W. Junk Publishers, The Hauge, The Netherlands, 1966.

91. M. A. Zahran and A. A. Abdel Wahid, "Contributions to the ecology of halophytes," Tasks for Vegetation Science, vol. 2, pp. 235–257, 1982.

92. A. M. Helalia, S. El-Amir, S. T. Abou-Zeid, and K. F. Zaghloul, "Bio-reclamation of saline-sodic soil by Amshot grass in Northern Egypt," Soil and Tillage Research, vol. 22, no. 1-2, pp. 109–115, 1992.

93. Z. Ke-Fu, "Desalinization of saline soils by Suaeda salsa," Plant and Soil, vol. 135, no. 2, pp. 303–305, 1991.

94. A. Hamidov, J. Beltrao, A. Neves, V. Khaydarova, and M. Khamidov, "Apocynum lancifolium andChenopodium album—potential species to remediate saline soils," WSEAS Transactions on Environment and Development, vol. 3, no. 7, pp. 123–128,

2007.

95. F. A. Nasir, "Bioreclamation of a saline sodic soil in a semi arid region/Jordan," American-Eurasian Journal of Agricultural and Environmental Sciences, vol. 5, pp. 701–706, 2009.
96. M. Rabhi, S. Ferchichi, J. Jouini et al., "Phytodesalination of a salt-affected soil with the halophyteSesuvium portulacastrum L. to arrange in advance the requirements for the successful growth of a glycophytic crop," Bioresource Technology, vol. 101, no. 17, pp. 6822–6828, 2010.
97. K. F. Zhao, W. J. Zhang, H. Fan, J. Song, and X. V. Jiang, "Biological measures in amelioration and utilization of saline soils," Soil Bulletin, vol. 32, pp. 115–119, 2001.
98. A. Graifenberg, L. Botrini, L. Giustiniani, F. Filippi, and M. Curadi, "Tomato growing in saline conditions with biodesalinating plants: Salsola soda L., and Portulaca oleracea L.," Acta Horticulturae, vol. 609, pp. 301–305, 2003.
99. M. Tester and R. Davenport, "Na^+ tolerance and Na^+ transport in higher plants," Annals of Botany, vol. 91, no. 5, pp. 503–527, 2003.
100. H. Kushiev, A. D. Noble, I. Abdullaev, and U. Toshbekov, "Remediation of abandoned saline soils using Glycyrrhiza glabra," International Journal of Agricultural Sustainability, vol. 3, no. 2, p. 102, 2005.
101. N. Al-Awadhi, F. K. Taha, and Petroleum Energy Center, New Technologies for Soil Reclamation and Desert Greenery: Proceedings of the Joint KISR-PEC Symposium, Kuwait Institute for Scientific Research, 2002.
102. S. M. Shiyab, R. A. Shibli, and M. M. Mohammad, "Influence of sodium chloride salt stress on growth and nutrient acquisition of sour orange in vitro," Journal of Plant Nutrition, vol. 26, no. 5, pp. 985–996, 2003.
103. E. R. de Souza, M. B. G. dos Santos Freire, K. P. V. da Cunha, C. W. A. do Nascimento, H. A. Ruiz, and C. M. Teixeira Lins, "Biomass, anatomical changes and osmotic potential in Atriplex nummulariaLindl. cultivated in sodic saline soil under water stress," Environmental and Experimental Botany, vol. 82, pp. 20–27, 2012.

104. USEPA, Government Report: Introduction to Phytoremediation, The U.S. Environmental Protection Agency, 2000.
105. M. Qadir and J. Oster, "Vegetative bioremediation of calcareous sodic soils: history, mechanisms, and evaluation," Irrigation Science, vol. 21, no. 3, pp. 91–101, 2002.
106. G. R. Sandhu and R. H. Qureshi, "Salt affected soils of Pakistan and their utilization," Reclamation & Revegetation Research, vol. 5, no. 1–3, pp. 105–113, 1986.
107. M. R. Chaudhry and M. Abaidullah, "Economics and effectiveness of biological and chemical methods in soil reclamation," Pakistan Journal of Agricultural Sciences, vol. 9, pp. 106–114, 1988.

Chapter 2

Effect of Foliar Application of Micronutrients on Fruit Quality of Peach

Amjad Ali[1], Sajida Perveen[2], Syed Noor Muhammad Shah[3,4], Zengqiang Zhang[1], Fazli Wahid[2], Mohib Shah[5], Shahida Bibi[6], and Abdul Majid[7]

[1]College of Natural Resources and Environment, Northwest A&F University, Yangling, China

[2]Soil and Environmental Sciences, University of Agriculture, Peshawar, Pakistan

[3]Department of Horticulture, Faculty of Agriculture, Gomal University, D. I. Khan, Pakistan

[4]College of Horticulture, Northwest A&F University, Yangling, China

[5]Department of Botany, Abdul Wali Khan University, Mardan, Pakistan

[6]Department of Weed Science, University of Agriculture, Peshawar, Pakistan

[7]Department of Agriculture Chemistry, Faculty of Agriculture, Gomal University, D.I. Khan, Pakistan

ABSTRACT

This experiment was conducted to study the effect of foliar spray of micronutrients on quality of peach fruits at Horticulture Farm, University of Agriculture Peshawar during 2010. The experiment was conducted in randomized complete design with six treatments and four replications on peach orchards (early grand). Our pre-treatment soil analysis showed silt loam soil class, alkaline and calcareous in nature and adequate in organic matter. The soil was deficient in P, Zn, Fe and B, whereas adequate in Mn and Cu. The fruit quality was evaluated and maximum fruit length, diameter and yield were noted in T_6 (Zn + Cu + Fe + Mn + B). The juice pH decreased and the juice acidity increased in a linear fashion after foliar spray of micronutrients. The total soluble solids of fresh fruit juice ranged 7.01% - 8.88% and vitamin C ranged from 4.80% - 7.90% after foliar spray. So the foliar spray of micronutrients significantly ($P \leq 0.05$) affected the quality of peach fruit.

INTRODUCTION

Peach (Prunus persical L.) belongs to the family Rosaceae with other species collectively known as "stone fruits". Important varieties grown worldwide are Cling and Free Stone. Peshawar and Swat are best suited for Early Grand, Florida King 6-A and 7, 8, 9 peach cultivars. The soil and climatic conditions of Khyber Pakhtunkhwa are suitable for the production of peach. According to Agricultural Statistics of Pakistan [1] , peach orchards covered 15.8 thousand ha area in Pakistan, producing 83.7 thousand tons of peach fruit, while in Khyber Pakhtunkhwa it covers an area of 6.2 thousand hectares producing 57.8 thousand tons of peaches. The main sources of micronutrients are parent material, sewage sludge, town refuse, fungicides and farmyard manure and are present in small amount in soil [2] . Micronutrients play a key role in plants growth and productivity [3] . Zinc is a component of almost 60 enzymes; it has a role in producing the growth hormone IAA. Zinc plays a key role in N metabolism of plant and Zn deficient plants have reduced protein content [4] [5] . Copper is essential for the growth of plant and helps in the formation of vitamin A [6] . Symptoms of copper deficiency include abnormal and stunted leaf development, stunted shoot growth with dieback, reduced flowering and fruit set, and small fruit with poor color and quality [7] . Iron forms important enzymes and is associated with chlorophyll synthesis. Iron deficient peach trees have small size, resulting in a large decrease in acceptable fruits [8] . Iron chlorosis reduces fruit production, reducing the number of fruits per tree, fruit size, fruit yield and affecting fruit quality parameters such as firmness and acidity. Fe fertilization increases fruit quality and yield in many crops [9]. Manganese is involved in photosynthesis, metabolism of nitrogen and carbohydrate. Boron is helpful in plant growth and productivity. It increases the germination of pollen grains and elongation of pollen tube, fruit setting and yield in orchards [10] . Boron is responsible for activation of dehydrogenase enzymes, sugar translocation, nucleic acids and plant hormones [11]. There are many soil and climatic factors affecting the availability of micronutrients to plants. The major factors include: pH, soil water content, organic matter, interaction among the nutrients and temperature. The availability of B and AB-DTPA extractable Zn, Cu, Fe and Mn decreases with increasing soil pH [12] [13]. Study shows the deficiency of Zn (17 %), Cu (13%), Fe (45%), Mn (2%) and B (95%) in orchards. The soils are alkaline, non-saline, calcareous and low in organic matter in Peshawar Valley [14]. These micronutrients are more soluble in the acidic pH and their availability in soils varies considerably with the seasonal changes in temperature and moisture [15] . As peach trees are deep rooted so foliar application of micronutrients is a valuable tool for fighting against the micronutrient deficiencies. Curing micronutrient deficiencies through foliar application is a common practice in getting profitable yield and good quality fruit [16]. Foliar spray of micronutrients is advantageous over soil application, because of rapid response, effectiveness and elimination of deficiency symptoms due to certain micronutrients. Results indicated that spraying the trees twice or thrice is more effective than spraying once a year in improving micronutrients content of peach trees [10]. This study was aimed to study the effect of micronutrients (Zn, Cu, Fe, Mn and B) spray on fruit set, yield and fruit quality of peach orchards.

MATERIALS AND METHODS

Experimental Design and Fertilizers: The experiment was conducted on peach orchards at University of Agriculture, Peshawar-Pakistan during 2010 (February-June).Twenty four trees were selected for this purpose. Six treatments with four replications were applied. A basal dose of 150 kg N ha^{-1}, 100 kg P_2O_5 ha^{-1} and 100 kg K_2O ha^{-1} as Urea, SSP and SOP was applied. The micronutrients Zn, Cu, Fe, Mn, B at rate of 5, 2.5, 5, 5, 1 Kg ha^{-1} respectively were applied in split dose. The detail of the micronutrients doses and their combinations is given in Table 1. Before incorporation of basal dose of fertilizers, composite soil sample was taken for the determination of various physico-chemical characteristics. After foliar spray the fruit length (cm), diameter (cm), No. of fruits $tree^{-1}$, yield $tree^{-1}$, fruit volume, Juice pH, Juice acidity, Total Soluble Solids and Vitamin C were recorded.

Soil Analysis: Soil samples were analyzed for soil texture [17], pH [18], EC [19], organic matter [20], lime [21] , extractable micronutrients, P and K [22] and B in leaf and soil [23] .

Fruit Analysis: The fruit analysis was performed at edible maturity stage. Physical properties were measured by Maurer and Taylor method [24]. The chemical properties of fruit were determined according to the methods of AOAC [25].

Statistical Analysis: Data were subjected to analysis of variance and differences among treatments were evaluated [26] [27]

RESULTS AND DISCUSSIONS

Physico-chemical characteristics of experimental soil: The Physico-chemical characteristics of soil are presented in Table 2. The soil was silt loam, pH ranged from 7.73 - 7.96, which was low in the surface and increased linearly with depth. The differences might be due to leaching of bases from the surface and its accumu

Table 1: Micronutrient treatments used for foliar spray on peach orchard

Treatments	Foliar spray of various micronutrients

T1	Control Zn (5 Kg ha–1)
T2	Zn + Cu (5 Kg ha–1 + 2.5 Kg ha–1)
T3	Zn + Cu + Fe (5 Kg ha–1 + 2.5 Kg ha–1 + 5 Kg ha–1)
T4	Zn + Cu + Fe + Mn (5 Kg ha–1+ 2.5 Kg ha–1 + 5 Kg ha–1+ 5 Kg ha–1)
T5	
T6	Zn + Cu + Fe + Mn + B (5 Kg ha–1+ 2.5 Kg ha–1+ 5 Kg ha–1+ 5 Kg ha–1+ 1 Kg ha–1)

Table 2: Physico-Chemical Characteristics of Experimental Soil

Properties of soil	Unit	0 - 15 cm	15 - 30 cm	30 - 45 cm	45 - 60 cm
pH (1:5)	-	7.73	7.89	7.93	7.96
EC(1:5)	dS m–1	0.20	0.20	0.17	0.16
Soil Organic Matter (%)	%	2.07	1.93	1.86	1.83
Lime (%)	%	20.75	19.25	17.50	17
AB-DTPA extractable P	mg Kg–1	6.47	6.28	5.52	5.39
AB-DTPA extractable K	mg Kg–1	100	100	50	50
AB-DTPA extractable Zn	µg g–1	1.14	1.00	0.57	0.56
AB-DTPA extractable Cu	µg g–1	5.38	5.30	3.29	2.52
AB-DTPA extractable Fe	µg g–1	3.76	3.28	2.45	2.29
AB-DTPA extractable Mn	µg g–1	19.16	18.67	17.50	14.50
Dilute HCl available B	µg g–1	0.45	0.29	0.32	0.37
Soil Texture	Silt loam				

lation in subsurface. The pH of Peshawar soil is generally alkaline [12]. The EC of the suspension varied from 0.16 - 0.20 dS m^{-1}. The results showed that the soil organic matter ranged 1.83% - 2.07 % (marginal-adequate). The soil is calcareous due to the high $CaCO_3$ content in the parent material [12]. The extractable P ranged 5.39 - 6.47 mg Kg^{-1}. Data showed that extractable P was marginal. The K concentration in the surface (0 – 30 cm) soil was marginal and low in the subsoil (30 - 60 cm).

Micronutrients content in soil of peach orchards: The data regarding the AB-DTPA extractable micronutrients is presented in Table 2. The Zn content ranged 0.56 - 1.14 µg g^{-1}. The concentration ranged from low (<0.9 µg g^{-1}) to marginal (0.9 - 1.5 µg g^{-1}) in soil. About 70% of the cultivated soils in Pakistan are deficient in Zn [28]. The Cu content was high (>0.5 µg g^{-1}) in soil samples collected from four depths. The Fe content ranged low (<3.0 µg g^{-1}) to marginal (3.0 - 5.0 µg g^{-1}) levels. The Mn content ranged 14.50 to 19.16 µg g^{-1} which was high (>1.0 µg g^{-1}) in all the depths. The soil of Khyber Pakhtunkhwa is adequate in Mn content as reported by Khattak and Perveen (1985). The acid soluble B ranged 0.37 - 0.45 µg g^{-1}. The concentration of B was low (<0.45 µg g^{-1}). Khattak and Perveen [29] reported 98% deficiency of B in soils of Khyber Pakhtunkhwa.

FRUIT QUALITY

Physical Properties of Fruit

Fruit Length (Cm): Fruit length was significantly ($P \leq 0.05$) influenced by micronutrients spray (Table 3). Maximum fruit length (6.41 cm) was noted in T_6 (Zn + Cu + Fe + Mn + B), followed by T_5 (Zn + Cu + Fe + Mn) and T_4 (Zn + Cu + Fe) whereas significantly short fruits (5.64 cm) were noted in control and Zn applied plants. These results were found in harmony with the work of Baloch [30] and Ramezaniand Shekafandeh [31]

Fruit Diameter (Cm): Fruit diameter was significantly ($P \leq 0.05$) increased by micronutrients spray (Table 3). Larger and healthy fruits (5.83 cm) were noted in T_6 (Zn + Cu + Fe + Mn + B) whereas, the poor were found in T_1 (control). Moreover, T_2 is similar to T_3. The increase in fruit diameter was linear from T_1 to T_6. The same trend was reported by Shahin et al. [32] and Ramezani and Shekafandeh [31] while working on plum and olive fruit, respectively.

Number of Fruit per Tree: The results significantly ($P \leq 0.05$) varied when different treatments of foliar spray were applied (Table 3). Moreover, maximum number of fruits (615) were obtained in T_6(Zn + Cu + Fe + Mn + B), followed by T_5 (Zn + Cu + Fe + Mn) with a fruit number of 585 followed by T_4 (Zn + Cu + Fe), while T_1 (control) gave minimum (448) number of fruits. Same results were reported by Shahin et al. [32] and Hassan et al. [4]

Fruit Yield (Kg/Tree): Fruit yield was significantly ($P \leq 0.05$) increased by micronutrients spray (Table 3). The higher yield (64.54 Kg/tree) was obtained in case of T_6 (Zn + Cu + Fe + Mn + B). The results regarding T_2 (Zn), T_3 (Zn + Cu), T_4 (Zn + Cu + Fe) and T_5 (Zn + Cu + Fe + Mn) were statistically comparable but better than the control (untreated) where minimum yield (40.62 Kg/tree) was obtained. These results were found in harmony with the work of Nasiriet al. [33] , Ibrahim et al. [34] and Khayyat et al. [35] , where micronutrients were applied to citrus and date palm, which increased the yield in a linear fashion.

Fruit Volume (Cm^3): The results showed significant ($P \leq 0.05$) increase in fruit volume (Table 3). The maximum fruit volume (267.51 cm^3) was obtained in T_6 (Zn + Cu + Fe + Mn + B) followed by T_5 (Zn + Cu + Fe + Mn) and T_4 (Zn + Cu + Fe). The poor fruit volume (195.01 cm^3) was noted in T_1 (control). The increase in fruit volume is due to availability of applied nutrients.

Chemical Properties of Fruits

Juice pH: It can be seen from the results that the application of foliar spray significantly ($P \leq 0.05$) decreased the juice pH (Table 4). It was noted that higher juice pH (3.86) was reported in T_1(control) followed by T_2 (Zn only), while the lower juice pH was noted in T_6, where all the micronutrients were applied. These findings were in line with the work of Tariq et al. [14]

Juice Acidity (%): The results showed that foliar spray significantly ($P \leq 0.05$) increased juice acidity over control (Table 4). It was noted that the maximum juice acidity (0.39%) was reported in plants sprayed with T_6 followed by T_5 and T_4. The minimum juice acidity (0.24%) was noted in control.

Total Soluble Solids (TSS): It can be seen from the results that the application of foliar spray significantly ($P \leq 0.05$) increased the total soluble solids (Table 4). It was noted that significantly ($P \leq 0.05$) higher total soluble solids (8.87%) were reported in T_6 followed by T_5 and T_4. The lower total soluble solids were noted in T_1 (control).

Vitamin C (mg/100mL): It can be seen from the results that the foliar spray significantly ($P \leq 0.05$) increased the vitamin C content in peach fruit (Table 4).

Table 3: Effect of foliar spray of micronutrients on the physical properties of fruit

Treatments	Length (cm)	Diameter (cm)	No. of fruit/tree	Fruit Yield Kg tree–1	Volume cm3
Control (T1)	5.64 c	5.19 d	488 e	40.62 c	195.01 f
Zn (T2)	5.71 c	5.41 c	525 d	52.57 b	206.26 e
Zn + Cu (T3)	6.01 b	5.48 c	541 cd	51.66 b	222.51 d
Zn + Cu + Fe (T4)	6.07 b	5.57 bc	557 c	52.89 b	236.26 c
Zn + Cu + Fe + Mn (T5)	6.22 ab	5.68 ab	585 b	51.80 b	248.76 b
Zn + Cu + Fe + Mn + B (T6)	6.41 a	5.83 a	615 a	64.54 a	267.51 a
LSD ($P \leq$ 0.05)	0.23	0.19	20.23	9.09	9.95
% CV	2.48	2.20	2.45	11.53	2.89

*Means are different significantly in columns at $P \leq 0.05$.

Table 4: Effect of foliar spray of micronutrients on the chemical properties of the fruits

Treatments	Juice pH	J. acidity %	TSS %	Vitamin C %
Control (T1)	3.86 a	0.24 d	7.01 e	4.80 f
Zn (T2)	3.78 a	0.25 cd	7.26 de	5.39 e
Zn + Cu (T3)	3.70 a	0.28 bc	7.58 cd	5.91 d
Zn + Cu + Fe (T4)	3.70 a	0.29 b	8.01 bc	6.62 c
Zn + Cu + Fe + Mn (T5)	3.68 a	0.31 b	8.36 b	7.14 b
Zn + Cu + Fe + Mn + B (T6)	3.31 b	0.39 a	8.88 a	7.90 a
LSD ($P \leq 0.05$)	0.33	0.04	0.43	0.46
% CV	5.97	8.27	3.65	4.83

*Means are different significantly in columns at $P \leq 0.05$.

It was noted that the maximum vitamin C (7.90 mg/100mL) content was reported in T_6, where all the micronutrients were applied. The minimum vitamin C (4.80 mg/100mL) content was reported in T_1 (control) where no spray was applied. These results revealed that foliar spray of micronutrients has a positive effect on vitamin C content. Hassan et al. [4] reported similar findings by applying micronutrients to plum orchards.

CONCLUSIONS

The obtained results in the present research proved that foliar application of micronutrients (Zn, Cu, Fe, Mn and B) alone and in combination were the most effective treatments in increasing chemical and physical parameters of peach fruit. To increase fruit length (cm), diameter (cm), no. of fruit/tree, fruit yield (kg/tree), volume (cm^3), juice (pH), juice acidity (%), TSS (%) and vitamin C (%) under silt loam, calcareous and alkaline condition of Peshawar soil, it is recommended to apply foliar spray twice a year to the orchards in order to enhance the efficiency of micronutrients and avoid losses by leaching and interaction among the nutrients.

ACKNOWLEDGEMENTS

We are thankful to Professor Dr. Sajida Perveen and Professor Dr. M. Jamal Khan Khattak, Department of Soil and Environmental Sciences, University of Agriculture, Peshawar, for their kind supervision and help in designing and executing the experiment. We are also thankful to the Department of Soil and Environmental Sciences, University of Agriculture, Peshawar for providing lab and research facilities.

REFERENCES

1. Pakistan, G. (2009) Agriculture Statistics of Pakistan 2008-09. Ministry of Food and Agriculture, Economic Wing, Islamabad.
2. Nafees, M., Jan, M.R., Khan, H., Rashid, N. and Khan, F.U. (2009) Soil Contamination in Swat Valley Caused by Cadmium and Copper. Sarhad Journal of Agriculture, 25, 37-43.
3. Bansal, O.P. (1992) Changes in Micronutrients (Cu, Zn, Mn, Ni and Cd) Concentration in Soil Amended with Sludge. Clay Research, 2, 29-32.
4. Hassan, H.S.A., Sarrwy, S.M.A. and Mostafa, E.A.M. (2010) Effect of Foliar Spraying with Liquid Organic Fertilizer, Some Micronutrients and Gibberellins on Leaf Minerals Content, Fruit Set, Yield, and Fruit Quality of "Hollywood" Plum Trees. Agriculture and Biology Journal of North America, 1, 638-643.
5. Mengel, K., Kosegarten, H., Kirkby, E.A. and Appel, T. (2001) Principles of Plant Nutrition. Springer, New York. http://dx.doi.org/10.1007/978-94-010-1009-2
6. Martens, D.C. and Westermann, D.T. (1991) Fertilizer Application for Correcting Micronutrient Deficiencies. Micronutrients in Agriculture, Soil Science Society of America, Madison, 549-592.

7. Stiles, W. C. and Reid, W.S. (1991) Orchard Nutrition Management. Cornell Cooperative Extension Information Bulletin, 219, 22.

8. Powell, J.C., Lyons, C.G. and Haby, V.A. (1995) Effects of Copper, Zinc, and Sulfur Application to Peach Trees on Coastal Plain Soil. Communications in Soil Science & Plant Analysis, 26, 1637-1648. http://dx.doi.org/10.1080/00103629509369396

9. Àlvarez-Fernàndez, A., Abadía, J. and Abadía, A. (2006) Iron Deficiency, Fruit Yield and Fruit Quality. Iron Nutrition in Plants and Rhizospheric Microorganisms, Springer Netherlands, 85-101.

10. El-Sheikh, M.H., Khafgy, S.A.A., and Zaied, S.S. (2007) Effect of Foliar Application with Some Micronutyrients on Leaf Mineral Content, Yield and Fruit Quality of "Florida Prince Desert Red" Peach Trees. Journal of Agricultural and Biological Science, 3, 309-315.

11. Brady, N.C. and Weil, R.R. (1996) The Nature and Properties of Soils. Prentice Hall Inc., New Jersey.

12. Rashid, M., Bhatti, A.U., Khan, F.U. and Wasiullah (2008) Physico-Chemical Properties and Fertility Status of Soils of District Peshawar and Charsadda. Soil and Environment, 27, 228-235.

13. Khattak, J.K., Sharif, M. and Naz, S. (1994) Nutrient Status of Citrus Orchard Soils in Peshawar Valley [Pakistan]. Sarhad Journal of Agriculture, 10, 451-460.

14. Tariq, M., Shah, Z. and Ali, A. (2008) Micronutrients Status of Plum Orchards in Peshawar Valley. Soil and Environment, 27, 223-227.

15. Hodgson, J.F. (1963) Chemistry of the Micronutrient Elements in Soils. In: Norman, A.G., Ed., Advances in Agronomy, Academic Press, Salt Lake City, 119-159.

16. Waheed, U.R. (1992) Effect of Foliar Application of Micronutrients in Combination with Urea on Yield and Fruit Quality of Sweet Oranges. M.Sc. (Hons) NWFP, Agriculture University Peshawar, Pakistan.

17. Bouyoucos, G.J. (1962) Hydrometer Method Improved for Making Particle Size Analyses of Soils. Agronomy Journal, 54, 464-465.http://dx.doi.org/10.2134/agronj1962.00021962005400050028x

18. Mclean, E.O. (1982) Soil pH and Lime Requirement. In: Page, A.L., Ed., Methods of Soil Analysis. Part 2. Chemical and Microbiological Properties, American Society of Agronomy, Soil Science Society of America, 199-224.
19. Richards, L.A. (1954) Diagnosis and Improvement of Saline and Alkali Soils. Soil Science, 78, 154. http://dx.doi.org/10.1097/00010694-195408000-00012
20. Nelson, D.W. and Sommers, L.E. (1996) Total Carbon, Organic Carbon, and Organic Matter. In: Sparks, D. L., Helmke, P.A., Loeppert, R.H., Soltanpour, P.N., Tabatabai, M.A., Johnston, C. and Sumner, M.E., Eds., Methods of Soil Analysis, Soil Science Society of America Inc., Madison, 961-1010.
21. Page, A.L. (1982) Methods of Soil Analysis. Part 2. Chemical and Microbiological Properties. American Society of Agronomy, Soil Science Society of America, Madison.
22. Soltanpour, P.N. and Schwab, A.P. (1977) A New Soil Test for Simultaneous Extraction of Macroand MicroNutrients in Alkaline Soils 1. Communications in Soil Science & Plant Analysis, 8, 195-207. http://dx.doi.org/10.1080/00103627709366714
23. Bingham, F.T. (1982) Boron. In: Page, A.L., Ed., Methods of soil Analysis Part-2 Chemical and Mineralogical Properties, American Society of Agronomy, Madison, 431-448.
24. Maurer, M.A. and Taylor, K.C. (1999) Effect of Foliar Boron Sprays on Yield and Fruit Quality of Navel Oranges. Cooperative Extension and Agricultural Experiment Station, University of Arizona, College of Agriculture.
25. AOAC (1990) Official Methods of Analysis of the AOAC. In: Helrich, K.C., Ed., Official Methods of Analysis of the AOAC, Association of Official Analytical Chemists Inc., Arlington, 1298.
26. STATISTIX (2000) STATISTIX7 User's Manual. Analytical Software, Tallahassee.
27. Scientific, J. (1991) Table Curve V. 3.0. User's Manual Version 3.0 AISN software. Jandel Scientific, Corte Madera.
28. Rashid, A. (1996) Secondary and Micronutrients. In: Bantel, E.S., Ed., Soil Science, National Book Foundation, Islamabad, 341-386.

29. Khattak, J.K. and Parveen, S. (1985,) Cooperative Research Programme on Micronutrient Status of Pakistan Soils. NWFP Agricultural University Peshawar, Pakistan, 52.

30. Baloch, Q.B., Chachar, Q.I. and Tareen, M.N. (2008) Effect of Foliar Application of Macro and Micro Nutrients on Production of Green Chilies (Capsicum annuum L.). Journal of Agriculture Technolgy, 4, 177-184.

31. Ramezani, S. and Shekafandeh, A. (2009) Roles of Gibberellic Acid and Zinc Sulphate in Increasing Size and Weight of Olive Fruit. African Journal of Biotechnology, 8, 6791-6794.

32. Shahin, M.F.M., Fawzi, M.I.F. and Kandil, E.A. (2010) Influence of Foliar Application of Some Nutrient (Fertifol Misr) and Gibberellic Acid on Fruit Set, Yield, Fruit Quality and Leaf Composition of "Anna" Apple Trees Grown in Sandy Soil. Journal of American Science, 6, 202-208.

33. Nasiri, Y., Zehtab Salmasi, S., Nasrullahzadeh, S., Najafi, N. and Ghassemi-Golezani, K. (2010) Effects of Foliar Application of Micronutrients (Fe and Zn) on Flower Yield and Essential Oil of Chamomile (Matricaria chamomilla L.). Journal of Medicinal Plant Research, 4, 1733-1737.

34. Ibrahim, M., Ahmad, N., Anwar, S. A. and Majeed, T. (2007) Boron Nutrition and Boron Application in Crops. In: Xu, F., Goldbach, H.E., Brown, P.H., Bell, R.W., Fujiwara, T., Hunt, C.D., Goldberg, S. and Shi, L., Ed., Advances in Plant and Animal Boron Nutrition, Springer Netherlands, 179-182.

35. Khayyat, M., Tafazoli, E., Eshghi, S., and Rajaee, S. (2007) Effect of Nitrogen, Boron, Potassium and Zinc Sprays on Yield and Fruit Quality of Date Palm. American-Eurasian Journal of Agricultural & Environmental Sciences, 2, 289 296.

Chapter 3

Impact of Addition of FGDB as a Soil Amendment on Physical and Chemical Properties of an Alkali Soil and Crop Yield of Maize in Northern China Coastal Plain

H.-L. Yu[1, 2, 3], W. Gu1, J. Tao[1], J.Y. Huang[3,4], and H.S. Lin[3]

[1]The State Key Laboratory of Earth Surface Process and Resource Ecology, Beijing Normal University, Beijing 100875, China

[2]College of Resources and Environment, Ningxia University, Yinchuan 750021, China

[3]College of Agricultural Sciences, the Pennsylvania State University, State College, PA 16803, USA

[4]Center of New Technology Application and Research, Ningxia University, Yinchuan 750021, China

ABSTRACT

To evaluate the effect of Flue gas desulfurization byproduct(FGDB)as a soil amendment on growth and yield of maize (Zea mays) and to determine the impact of FGDB additions on soil fertility characteristics in alkaline clayey soils, a 2-year field experiment was conducted in Huanghua, in Northern China Coastal Plain. The experiment included five treatments in which the soil was amended with FGDB at 15 cm depth at the rates of 0 $t{\cdot}hm^{-2}$, 4.50 $t{\cdot}hm^{-2}$, 9.00 $t{\cdot}hm^{-2}$, 13.5 $t{\cdot}hm^{-2}$, and 18.00 $t{\cdot}hm^{-2}$, respectively, before maize was planted. The values of soil pH, exchangeable sodium percentage (ESP), and bulk density (BD) of the soil decreased; however, values of electrical conductivity (EC), water holding capacity (WHC), and plant nutrients increased with FGDB application in the soil. Crop plants grow more readily in FGDB amended soils because of improved soil properties. The best ameliorative effect was obtained at the rate of 13.5 $t{\cdot}hm^{-2}$. The germination percentage, plant height, and crop yield successively increased in both years. The results indicated FGDB was an effective soil amendment for improving the physicochemical properties and nutrient balance, and enhancing crop germination, growth, and yield, particularly when applied at a suitable application rate.

INTRODUCTION

Salinization and alkalinization of soil are a widespread environmental problem. Alkaline-saline soils are widely distributed across the world, covering an area of 9.5×10^6 km^2, accounting for 7.3% of the Earth's land surface [1]. About 23% of the world's currently cultivated land area (1.5×10^9 ha) is saline and another 37% is alkaline. Soil degradation caused by salinization and alkalinization is of universal concern. In the northwestern, northern, northeastern, and coastal area of China, there are 346,000 km^2 saline and alkali soil, 1/3 of which are alkali soil (92,000 km^2 or 22.7 million acres). Alkalization and salinization of soils are important factors limiting agricultural productivity [2]. These soils are unproductive, impermeable, and hard due to the presence of undesirable salts on the surface.

The salinization and alkalinization of the soil in the coastal area of China are very serious. Crop yields in Bo Sea coastal plain of China are limited by the high soil alkalinity, which is common to this region. The chief characteristic of alkaline soils from agricultural standpoint is that they contain excessive quantities of exchangeable sodium and appreciable amounts of soluble sodium carbonate to adversely affect the growth of most crop plants. Alkali soils are characterized by exchangeable Na^+ (more than 15%), low quantities of Ca^{2+}, and pH values that usually range between 8.5 and 10.5. Excess exchangeable sodium has an adverse effect on the physical and nutritional properties of the soil. Plant growth in alkaline soils is adversely affected chiefly by the dispersive effect of excess exchangeable sodium resulting in poor physical soil properties and the effect of high soil pH on nutritional imbalance including a deficiency of calcium. Due to the poor physical properties, most alkaline clayey soil belongs to the middle and low-yield soils. Several techniques have been developed to improve the physical conditions, such as soil amendment using organic manures and industrial wastes. Improvement of alkaline soils essentially requires the replacement of sodium in the soil exchange complex by calcium through use of soil amendments and leaching and drainage of salts resulting from reaction of amendments with exchangeable sodium.

Saline wastelands have been successfully reclaimed for corn growth in Bo Sea coastal plain of China in recent years [3]. The primary management strategies adopted to accomplish this include the application of FGDB amendment on saline wasteland, which combines industrial waste recycling with agricultural soil improvement [4]. In this case, more FGDB requirement should be applied to replace the exchangeable Na^+required to neutralize all the soluble sodium carbonate in the soil.

Flue gas desulfurization byproducts (FGDBs) are often marketed as a soil conditioner for improving soil tillage. Considerable amounts of coal combustion products are generated when coal is burned for generation of electricity. In China about 80% of electricity is generated by coal-based thermal power plant, which produces more than 1000 mt FGDB per annum as a byproduct that creates the problem of its ecofriendly disposal. Most coal combustion products in China are presently discarded, especially in landfills, and landfill sites are becoming more limited and disposal costs continue to increase. To meet Clean Air Standards, wet scrubber technology currently used removes much of the SO_2 to form large amounts of flue gas desulfurization byproducts (FGDBs) [5, 6]. However, FGDB is also known as a potential environmental pollutant at both regional and global levels [7]. Thus, the use of FGDB as substitutes for agricultural limestone represents a potential beneficial use alternative to landfill disposal of these materials.

Many experiments and studies on the effect and potentiality of FGDB as an amendment in agricultural applications have been conducted by various agencies and research institutes at dispersed locations all over the world [8–12]. Its use in agriculture is mainly based on its liming potential and supply of nutrients which promote growth of plants and alleviate the condition of nutrient deficiency in soils [13–16]. The main component of FGDB is calcium sulfate ($CaSO_4$). However, FGDB is a useful ameliorant that may improve the physical, chemical, and biological properties of problem soils and is a source of readily available plant macro- and micronutrients [17]. Compared to most other calcium-rich soil amendments, such as limestone, FGDB is relatively soluble in water, dissolving up to 2 g per liter [18, 19]. The solubility of FGDB, when either incorporated or surface applied, permits a quick release of calcium (Ca^{2+}) and sulfate () ions into the soil solution. The supply of dissolved salt and Ca^{2+} ions, in particular, may reduce soil crusting and otherwise benefit soil structure. The aggregation of clay particles that help to form and stabilize soil structure is clearly enhanced by the presence of calcium on clay exchange sites.

Many researchers have already proposed that relatively low dose of FGDB as soil amendment improves the physical and chemical properties of the soil, for example, soil pH, bulk density, texture, water holding capacity, electrical conductivity, and so on [20]. As a consequence, the growth and yield of crops also were increased significantly.

It has been suggested that FGDB could be a useful amendment for alkaline soils and has positive effects on soil improvement and crop production because of high concentrations of sorbent components and significant neutralization potential [21–23]. For example, beneficial agricultural uses of FGDB application as amendment to alkaline soil to mitigate high pH problems provide plant nutrients, improve soil physical properties, help alleviate soil compaction, and improve aggregate stability of alkaline soils [24, 25].

There is no clear guidance with regard to the application rate of FGDB as soil amendment. In order to make efficient use of FGDB to enhance crop production, it is necessary to be able to determine the amount of FGDB that should be applied.

Keeping all these in mind, the present experiment has been designed to investigate the impact of the different application rates of FGDB amendments on the soil properties and the growth and yield of maize on the alkaline soils of the coastal wasteland. The specific objectives were to (1) determine the efficacy of FGDB ash as a soil amendment; (2) evaluate FGDB ash as a source of Ca and other nutrients for crops; (3) understand the feasibility of FGDB in the alkaline waste land reclamation without affecting the production of maize;(4) determine the appropriate FGDB application rate that should be applied on condition of certain soil and crop. Due to the increasing solid waste under Chinese scenario, the management of FGDB as soil amendment can give us a better reference for the better disposal of solid waste.

MATERIALS AND METHODS

Experimental Site

The field experiments were carried out during 2011 and 2012 in Huanghua, Hebei, China. The station (38°22′N, 117°21′E) is in the eastern part of the North China Plain. The climate of the area is temperate monsoon. Because of monsoon influence, rainfall is highly variable in this region. Annual average precipitation is 627 mm, of which 70–80% is during July–September [26]. The average evaporation in this region is 1989 mm, up to 4-5 times the annual average precipitation. Annual average temperature is 12.1°C. Seasonal drought is frequent in spring and autumn, and 90% of springs experience drought; the soluble salts accumulate in the upper soil profile. The groundwater level changes from 1 to 3.5 m. The dominant soils in the coastal plain are characterized by having high pH values and alkalinity, being clay-rich, being of slow permeability class, and being dominated by clay with less sand and silt fractions. The major salts are Nacl and $NaHCO_3$

[27].

Before treatments, surface soil samples at two different depths (0–10 cm and 10–20 cm) were collected, air-dried, and analyzed to determine fertility status, pH, FGDB requirements, and concentrations of elements essential for growth of higher plants. Some selected soil physical, chemical, and mineralogical properties are listed in Tables 1, 2, and 3.

Table 1: Basic soil properties of the initial soil profile

Soil layers	Soil mechanical composition (%)			ECe (ms/m)	pH	Soil bulk density (g cm −3)	Soil texture	Soil color
	<0. 002 mm	0.002–0.05 mm	0.05–2 mm					
0–10 cm	57.37	34.12	8.51	0.436	8.74	1.35	Clay	Dirty white
10–20 cm	64.66	31.23	4.11	0.296	8.91	1.43	Clay	Grayish

Values are average of triplicate measurements.

Table 2: Basic chemical properties of the initial alkaline soil

Parameter	Value
Organic matter (%)	1.41 ± 0.09
pH	8.59 ± 0.14
EC (mMho cm−1)	0.43 ± 0.003
Total N (%)	0.23 ± 0.004
Available P (cmol kg−1 soil)	4.16 ± 0.06
K (cmol kg−1 soil)	0.110 ± 0.017
Na (cmol kg−1 soil)	3.32 ± 0.02
Ca (cmol kg−1 soil)	1.26 ± 0.06
Mg (cmol kg−1 soil)	1.02 ± 0.05
SO4 (cmol kg−1 soil)	3.67 ± 0.05
ESP (%)	16.30 ± 0.16

Values are mean ± standard deviation of triplicate measurements.

Table 3: Basic mineralogical properties of the initial alkaline soil

Parameter	Content (%)
Primary mineral	
Quartz	9.2 ± 0.25
Mica	9.2 ± 0.25
Feldspar	9.2 ± 0.25
Clay mineral	
Smectite	9.2 ± 0.25
Illite	9.2 ± 0.25
Potash feldspar	9.2 ± 0.25
Albite	9.2 ± 0.25
Kaolinite	9.2 ± 0.25

Values are mean ± standard deviation of triplicate measurements.

Experimental Design

The FGDB (pH = 7.2) was obtained from Dagang Electricity Plant, Tianjin. Some selected physical, chemical, and mineralogical properties of FGDB are listed in Tables 4 and 5. The crop species used in the experiments was maize (Zea mays). The experiment field was cleared of any weeds and senesced plant material before the establishment of plots. Plot sizes were designed by 6 m × 8 m each and were arranged in a randomized block design with five replications. The two crops were planted in 40 cm rows at a population of 400 seeds per plot on April 20, 2011. In order to find the optimum FGDB application rate for maize in Bo Sea coastal plain of China, we set up the five FGDB application rates treatments in this study. The amount of FGDB required for the field trials was determined from treatment application rates and the plot areas for each treatment. Before seeding, the FGDB were applied at rates equivalent to $0\,t{\cdot}hm^{-2}$, $4.50\,t{\cdot}hm^{-2}$, $9.00\,t{\cdot}hm^{-2}$, $13.5\,t{\cdot}hm^{-2}$, and $18.00\,t{\cdot}hm^{-2}$, respectively, hereafter denoted as T0, T1, T2, T3, and T4. T0 is the unamended control treatment. FGDB treatments were applied to the crops plots at 15 cm depth by filling a wheelbarrow with FGDB, transporting the wheelbarrow load to the plot and distributing the FGDB evenly over the plot area using small plastic pails. In 2012, the applications were repeated in the same way.

Table 4: Some mineralogical and physical properties of FGDB used to amend alkaline soil

Mineral component	Water content (%)	CaSO4 · 2H2O %	Particle size %			Texture	Insoluble residue (%)
			<30 μm	30–60 μm	>60 μm		
Gypsum and quarts	10	99	3.2	0.14	96.66	Silt sized	1

Average of triplicate measurements.

Table 5: Some chemical properties of FGDB used to amend alkaline soil

Components	SO3	CaO	MgO	Al2 O3	Cl	Na2O	K2O	P2O5	Heavy metals
Contents	48. 52	34. 12	2.11	1. 00	0. 57	0.27	0.11	1.12	1.08

Average of triplicate measurements.

Plant Management and Measurements

Plant seedlings productivity was recorded on the 15th day after seeding. Weeds were controlled by postemergence application of herbicide. Seed germination is usually used to screen crops for their tolerances to soil alkalinity. This experiment was performed to evaluate the effects on the germination rate of seeds in soil amended with FGDB. 400 seedlings of maize were planted in each plot with different application rates of FGDB amended soils. A total of 50 plots were used in this experiment.

Seedlings were monitored by calculating seed germination percentages 2 weeks after planting. At day 15, all the plants have germinated and the percentage of germination was counted. Plant height was measured by elongating the plant and measuring to the top of the upper leaf with a ruler at day 100 of each year. At day 180 of each year, plots were harvested. Yield was measured by open-air drying for 30 days; after that the dry weights were measured gravimetrically.

Soil Sampling and Analysis

To determine the original soil properties, three composite soil samples were collected from the entire experimental area at two depths (0–10 cm and 10–20 cm) prior to treatment establishment, and properties are reported in Table 1. Soil cores were obtained from each plot using an auger (2.0 cm in diameter and 15 cm high) on April 2 (before seeding) and October 6 (after harvest) in 2011 and on April 5 (before seeding) and October 10 (after harvest) in 2012. The samples were obtained from two different depths, that is, 0–10 and 10–20 cm deep. The three replicate soil samples were mixed into one sample per treatment. All soil samples were air-dried and passed through a 1 mm sieve. Soluble salt estimates and soil pH were based on extracts of saturated soil. Soil pH, electrical conductivity (EC), organic matter, cation exchange capacity (CEC), and texture were determined by using standard laboratory procedures. Water holding capacity (WHC) and bulk density (BD) were measured by the methods of Black [28]. Exchangeable Na^+, K^+, Mg^{2+}, and Ca^{2+} were extracted in ammonium acetate by repeated leaching methodology and the extracts were kept separately for their determination [29]. The content of N was assessed following modified Kjeldahl method and P and K were analyzed by spectrophotometer and flame photometer after acid digestion. The concentrations of heavy metals were obtained by EPA method 3050 [30]. In this study, ECe is the abbreviation of electrical conductivity of saturated-soil extract. Cation exchange capacity (CEC) and exchangeable sodium percentage (ESP) were calculated with the help of following equations:

$$CEC = \sum (Mg, K, Na, Ca, NH_4, H, Al),$$
$$EDP = \frac{\text{Exchangeable Na}}{\text{cation exchange capacity}} * 100. \quad (1)$$

The ECe (electrical conductivity of saturated-soil extract) of the soil at a depth of 20 cm is 0.453 mS/m (Table1), which is high for crops. The color of alkaline soil was dirty white to grayish and was of clay-type soil texture. The pH of the tested soil ranged from 8.74 to 8.91. The water holding capacity was 45%. The exchangeable Ca^{2+} and Na^{+} were 0.35 and 0.78%, respectively. Some selected soil properties such as soil texture, soil bulk density, ECe, and pH are listed in Table 1.

Statistical Analysis

The results obtained for each of the dependent variables in this study were analyzed statistically using a model that included treatment and replication as independent variables. All the statistical tests were performed using SPSS software (SPSS Inc., version 16.0).

RESULTS

Chemical and Physical Properties of FGDB

The result showed that FGDB used in the present study was relatively alkaline (pH = 7.5) and rich in soluble salts. The FGDB used in the present contained 10% free water and had a loose and uniform structure and a particle size of 30–60 μm (Table 4). Similar to natural gypsum, the major component is calcium sulfate ($CaSO_4$); additional components include Na_2O, K_2O, MgO, Cl, P_2O_5, and heavy metals [31, 32]. Chemical analyses of the FGDB collected in this study showed that heavy metals were presented at low concentrations (Table 5). The metal contents were much lower than concentration limits identified by government regulations for land application [33]. FGDB has a low bulk density, high surface area, and light texture [17, 34]. Some selected properties of the FGDB are provided in Tables 4 and 5.

FGDB to Improve Soil Physical Properties

Particle Size Analysis

The soil at the Huanghua research site is characterized by being clay-rich and having high soil pH value and slow permeability. The texture analysis revealed that the experimental soil had clay texture having 57.37% clay, 34.12% silt, and 8.51% sand, respectively. Greater silt contents are observed at the lower depths in FGDB amended soils (Table 6). Clay content does not vary greatly between deeper horizons. Differences in particle size analysis among three horizons result from changes in the silt and clay fractions. At the same time, improved soil texture of the upper layer of FGDB amended soils result from application of silt-sized FGDB. Application of FGDB can reduce clay dispersion and promote flocculation of soils. Results of particle size analysis (Table 6) showed that clay fractions dominated the soil texture for the experimental field. The USDA textural grouping was clayey for the two depths. All the four FGDB application rates showed the potential of improving soil texture. Application of higher rates of T3 and T4 can change the surface texture of soils, by increasing the silt content. As would be expected, the coarser fractions (sands and coarse silt) increased and the finer fractions decreased with depth.

Table 6: Soil texture of experimental plots (USDA classification)

Size fraction for depth interval	0–10 cm				10–20 cm			
	Sand (%)	Silt (%)	Clay (%)	Increase over silt (silt 34.12%)	Sand (%)	Silt (%)	Clay (%)	Increase over silt (silt 31.23%)
2011								
T0	8.51	34.12	57.37	0	4.11	31.23	64.66	0
TI	8.64	37.15	54.21	3.03	4.53	43.12	52.35	11.89
T2	8.53	38.44	53.03	4.32	4.62	45.64	49.74	14.41
T3	8.47	40.13	51.4	6.01	5.84	46.87	47.29	15.64
T4	8.54	41.62	49.84	7.5	6.12	48.45	45.43	17.22
2012								
T0	8.53	36.34	55.13	2.22	4.33	36.23	59.44	5

TI	8.79	39.46	51.75	5.34	4.53	47.12	48.35	15.89
T2	8.83	39.64	51.53	5.52	5.32	49.64	45.04	18.41
T3	8.56	44.83	46.61	10.71	6.15	49.96	43.89	18.73
T4	8.69	44.78	46.53	10.66	6.89	51.46	41.65	20.23

Average of triplicate measurements.

In both years, silt percentage due to FGDB application showed a significant increase over control. In 2011, a trend was observed in the surface soil layer (0–10 cm depth) with increasing dosage of FGDB an increase over control values for silt, from 37.15, 38.44, 40.13, and 41.62 at the application rates of T1, T2, T3, and T4, respectively. And a similar trend was observed at the deeper soil layer (10–20 cm depth) with increasing dosage of FGDB an increase from control values for silt, from 43.12, 45.64, 46.87, and 48.45 at the application rates of T1, T2, T3, and T4, respectively. In 2012, similarly, silt content showed an increase trend in both layers at all application rates. But the optimum increments were observed at T3 treatment unlike in 2011. It indicated that continuous high application rate of FGDB may have adverse effect on soil texture.

Bulk Density (BD) and Water Holding Capacity (WHC)

In both years, we collected soil sample after harvest and found that all bulk densities were significantly lower for the FGDB treatments at all the four application rates compared with bulk densities of the control. Even though the statistical report indicated a marked difference between control and other application rates, no significant difference was reported between T3 and T4, suggesting that at or over the rate T3 has a similar effect. Hence, the optimum rate of FGDB application to improve soil bulk density was 13.5 t hm^{-2}. In the control, the relatively high soil BD is consistent with the observed clay high fraction, low sand fraction, and exceptionally low OM content (Table 7). This condition is typical for local alkaline clay textured soils, which have potentially low infiltration, decreased aggregation, and lower water holding capacities.

Table 7: Treatment effect on surface soil (0–10 cm depth) BD (bulk density), WHC (water holding capacity), and OM (organic matter) in 2011 and 2012

	Year	T0	T1	T2	T3	T4
BD (g cm−3)	2011	1.37	1.34	1.32	1.24	1.19
	2012	1.43	1.31	1.29	1.18	1.16
WHC (%)	2011	46.1	49.9	53.6	64.9	66.3
	2012	47.4	51.3	63.2	71.2	75.3
OM (%)	2011	1.41	1.53	1.56	1.52	1.49
	2012	1.43	1.55	1.70	1.56	1.69

Average of triplicate measurements.

To test the effect of FGDB on the soils' water holding capacity (WHC), WHC of the unamended soils and soils treated with FGDB was determined. In comparison with the control soil, the 4.50 t hm^{-2}, 9.00 t hm^{-2}, 13.5 t hm^{-2}, and 18.00 t hm^{-2} FGDB amended soils showed a 8.24, 16.26, 40.78, and 43.81% higher water holding capacity in 2011, respectively. Similarly, in 2012, there was another increase of 7.6, 25, 33.42, and 37.05%, respectively, compared with the control. However, the rate of increase is slowed after 13.5 t hm^{-2}. For all treatments, increasing the application rate from 4.5 to 18.00 t hm^{-2} increased the WHC, whereas increasing the application rate to 13.5 t hm^{-2} brought WHCs up to values that are higher than at any of the other application rates, which was suggesting that the optimum rate of FGDB application to improve WHC was 13.5 t hm^{-2}.

FGDB to Improve Soil Chemical Properties

Soil pH

The soil pH data from plots established on this initially very alkaline soil demonstrates the significant alkaline neutralizing capacity of FGDB. The effect of modifying soil pH by FGDB as amendment showed persistent changes with respect to the rate of application and duration of experiment. Compared with the control, a dose-dependent and progressive decrease in soil pH was observed. The soil pH values, however, were significantly affected when the soil was amended with FGDB. At 0–10 cm depth, within the first year of application, soil pH decreased from 8.6 for the control to the maximum of 7.74 at T3 treatment at rate of 13.5 t hm^{-2}, which is considered suitable for nutrient availability and crop growth (pH 7.5–8.5). In terms of soil pH adjustment, apart from initially higher alkalinity in the FGDB treatments, the target soil pH range (pH 7.5–8.5) was achieved within the first year in all four FGDB application rates (Figure 1). And in 2012, the second year, a continuous decrease in soil pH was observed in all FGDB amended treatments, similarly with the biggest decrease occurring in T3. And no further increases in pH were observed for subsoil. The data indicated that the effect of modifying soil pH occurred primarily within the treatment zone. Results obtained in this study are consistent with the large body of literature which indicates that FGDB can act as an effective liming agent to neutralize solution pH and reduce exchangeable alkalinity values.

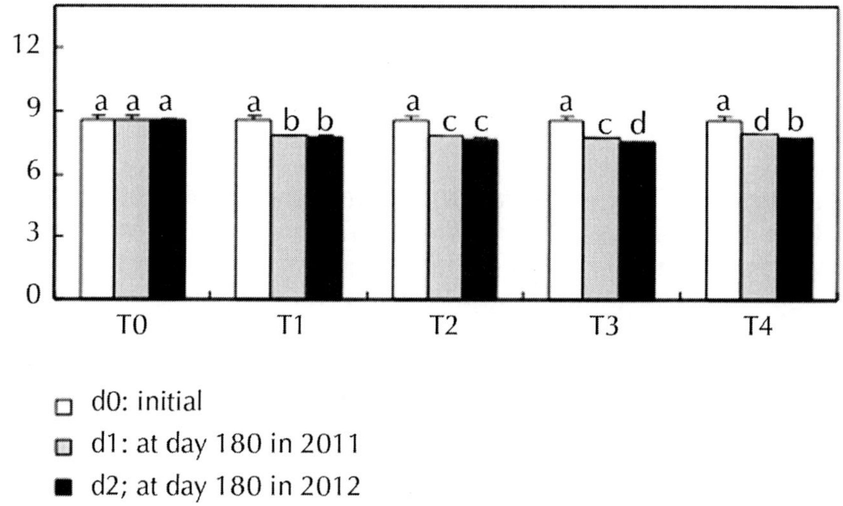

Figure 1: Effects of FGDB application on soil pH values for 2011 and 2012 (mean ± standard deviation of triplicate measurements. T0: control, T1: 4.50 t·hm^{-2}, T2: 9.00 t·hm^{-2}, T3: 13.50 t·hm^{-2}, and T4: 18.00 t·hm^{-2}. Lowercase letters denote statistically significant differences across treatments ($P \leq 0.05$) in 2011 and 2012, respectively.

EC

It is obvious that, after applying FGDB in alkaline clayey soil, soluble salts in FGDB were dispersed into the soils to a certain extent at the end of the experiment in 2011. Within the first year of application, at 0–10 cm depth, EC increased in response to FGDB addition. The average EC values in all the four treatments were higher than those of the control. In 2011, EC values for soils amended with FGDB increased from 0.44 to 0.46 mMho cm^{-1} at the T1 rate, to 0.49 mMho cm^{-1} at the T2 rate, to 0.51 mMho cm^{-1} at the T3 rate, and to 0.53 mMho cm^{-1} at the T4 rate, respectively (Table 8). When compared to the initial ECe value in 2011, the average value in 2012 was decreased to 0.41, 0.44, 0.46, and 0.51 mMho cm^{-1} for T1, T2, T3, and T4, respectively. Due to the bigger precipitation in 2012, more soluble salts leached into deeper soil profile resulting in the decrease of EC values compared with that of 2011. EC values of 0.5 to 1.0 can reduce germination and cause burning or wilting in some sensitive crops [35]. So, higher application rate of FGDB may influence potential growth of crops

Table 8: Selected chemical properties and nutrient (cmol kg^{-1}) concentrations of different FGDB amended soils

		T0	T1	T2		
EC (mMho cm−1)	d0	0.43 ± 0.003a	0.43 ± 0.003a	0.43 ± 0.003a	0.43 ± 0.003a	0.43 ± 0.003a
	d1	0.44 ± 0.002a	0.46 ± 0.003a	0.49 ± 0.003b	0.51 ± 0.002c	0.53 ± 0.005d
	d2	0.46 ± 0.003a	0.47 ± 0.004a	0.51 ± 0.005b	0.53 ± 0.004c	0.55 ± 0.006d
Total N (%)	d0	0.23 ± 0.004a	0.23 ± 0.004a	0.23 ± 0.004a	0.23 ± 0.004a	0.23 ± 0.004a
	d1	0.26 ± 0.004a	0.29 ± 0.005b	0.33 ± 0.002c	0.34 ± 0.004c	0.37 ± 0.002d
	d2	0.27 ± 0.003a	0.31 ± 0.001b	0.35 ± 0.004c	0.36 ± 0.002c	0.40 ± 0.004d
Available P (cmol kg−1 soil)	d0	4.16 ± 0.06a	4.16 ± 0.06a	4.16 ± 0.06a	4.16 ± 0.06a	4.16 ± 0.06a
	d1	4.64 ± 0.05a	5.61 ± 0.07b	5.92 ± 0.04c	6.42 ± 0.06d	6.55 ± 0.02d
	d2	4.87 ± 0.10a	5.85 ± 0.05b	6.10 ± 0.06c	6.56 ± 0.05d	6.61 ± 0.07d

K (cmol kg−1 soil)	d0	0.110 ± 0.017a	0.114 ± 0.012a	0.107 ± 0.015a	0.108 ± 0.013a	0.113 ± 0.010a
	d1	0.102 ± 0.008a	0.049 ± 0.009c	0.068 ± 0.010b	0.051 ± 0.008c	0.071 ± 0.003b
	d2	0.097 ± 0.007a	0.024 ± 0.003c	0.033 ± 0.010c	0.032 ± 0.002c	0.047 ± 0.007b
Na (cmol kg−1 soil)	d0	3.32 ± 0.02a	3.28 ± 0.07a	3.35 ± 0.09a	3.34 ± 0.04a	3.29 ± 0.09a
	d1	3.25 ± 0.06a	2.41 ± 0.08b	1.85 ± 0.05d	2.19 ± 0.01c	2.46 ± 0.08b
	d2	3.17 ± 0.07a	1.32 ± 0.04c	1.24 ± 0.06d	1.53 ± 0.02b	1.50 ± 0.09b
Ca (cmol kg−1 soil)	d0	1.26 ± 0.06a	1.32 ± 0.09a	1.27 ± 0.05a	1.23 ± 0.04a	1.30 ± 0.07a
	d1	1.31 ± 0.09e	2.05 ± 0.11d	4.11 ± 0.24c	5.93 ± 0.53b	11.75 ± 0.79a
	d2	1.19 ± 0.05e	1.82 ± 0.09d	3.66 ± 0.18c	4.89 ± 0.30b	8.27 ± 0.67a
Mg (cmol kg−1 soil)	d0	1.02 ± 0.05a	1.06 ± 0.08a	0.99 ± 0.04a	1.01 ± 0.05a	1.04 ± 0.03a
	d1	1.03 ± 0.02c	1.83 ± 0.01b	2.01 ± 0.04a	2.03 ± 0.05a	2.10 ± 0.07a
	d2	0.95 ± 0.03b	1.42 ± 0.04a	1.47 ± 0.05a	1.53 ± 0.03a	1.55 ± 0.06a
SO4 (cmol kg−1 soil)	d0	3.67 ± 0.05a	3.58 ± 0.04a	3.71 ± 0.02a	3.60 ± 0.04a	3.69 ± 0.03a
	d1	3.56 ± 0.04e	4.52 ± 0.07d	7.71 ± 0.08c	15.93 ± 0.14b	18.04 ± 0.13a
	d2	3.48 ± 0.04d	3.84 ± 0.05d	6.03 ± 0.05c	13.36 ± 0.10b	15.69 ± 0.09a
ESP (%)	d0	16.30 ± 0.16a	14.28 ± 0.18a	14.37 ± 0.17a	14.28 ± 0.23a	14.35 ± 0.14a
	d1	16.47 ± 0.20a	8.83 ± 0.10c	8.50 ± 0.11c	10.64 ± 0.15b	10.58 ± 0.18b
	d2	14.26 ± 0.15a	6.08 ± 0.08d	6.01 ± 0.13d	8.34 ± 0.10b	7.72 ± 0.05c

d0: initial; d1: at day 180 in 2011; d2: at day 180 in 2012.

Values = mean ± standard deviation of triplicate measurements.

Letters denote statistically significant differences across treatments (p=0.05).

ESP

Due to the excess exchangeable sodium, the initial ESP of the test soil is very high, 16.34. Soluble salts in FGDB were dispersed into soil and the exchangeable sodium was replaced by soluble calcium. Other substances in FGDB, such as sulphuric acid and sulphur, indirectly through chemical or biological action make the relatively insoluble calcium carbonate which commonly existed in alkaline soils available for replacement of sodium. The maximum exchangeable sodium percentage decrease was in case of T3, that is, 8.48% followed by T4 (9.40%), T2 (10.24%), and T1 (11.56%). As a result, the concentration of Na decreased, and the ESP values decrease synchronously with the decrease of soil pH values (Table 8). Compared with the control, supply of inherent calcium from the FGDB is likely to further accelerate the decrease process of ESP.

FGDB to Improve Plant Nutrients

Addition of FGDB to soil had some positive effect on nutrients status and other chemical parameters. Chemically, FGDB contains some growth-essential elements like K, Ca, and S which promote plant growth and also alleviate the condition of nutrient deficient soils. As calcium sulfate ($CaSO_4$) is the major constituent of FGDB, additional components, including Na_2O, K_2O, MgO, Cl, and P_2O_5, contribute largely to the nutrient resources of the FGDB amended soils. The soil was poor in organic C, N, and P contents. The addition of FGDB profoundly affected the cation and anion concentrations (Table 8). Similar progressive changes with respect to the rate of application were observed. Concentrations of calcium, sodium, and available phosphorus increased with respect to higher FGDB application. However, organic carbon, nitrate nitrogen, and potassium contents of soil decreased with the higher amendments of FGDB. The available P contents of the soils increased with an increase in the quantity of FGDB application due to the relatively high content of available P in the FGDB. The FGDB contained about 1120 mg kg^{-1} of available P_2O_5 (Table 5). Around 50.40, 100.8, 151.2, and 201.6 kg hm^{-2} of available P_2O_5 were supplied by the application of T1, T2, T3, and T4, respectively. Usually, FGDB contains a small quantity of organic matter and N; thus, there was an insignificant increase in available N in all the four FGDB amended soils in comparison to the control soil.

Concentrations of Cl^-, SO_4^{2+}, Ca^{2+}, and K^+ were greater in the FGDB amended soils compared with the control. Ca^{2+} concentrations in 2012 were greater in FGDB amended soils compared to those of 2011, reflecting a successive increase in Ca^{2+} concentration in soil amended with FGDB for 2 years compared with the control soil and the soil amended with FGDB for 1 year (Table 8). Similarly, total N contents increased progressively by up to 60% in 2011 and 73% in 2012, respectively, in response to FGDB addition. Other soil chemical parameters were not affected by FGDB addition. FGDB has both the soil amending and nutrient-enriching properties and is helpful in improving crop growth and yield in low fertility alkaline clay-rich soils.

The Effects on the Germination Percentage, Plant Height, and Crop Yield

Germination Percentage

Seedlings were monitored by calculating seed germination percentages 2 weeks after planting. At day 15, all the plant has germinated and counted the percentage of germination. Germination percentage was higher under all the four different FGDB treatments as compared to the control (Table 9). Germination percentage was maximum in treatment T3 in both years. It was 79.02% in 2011 and 91.25% in 2012, respectively. It was significantly more as compared to control. Treatments T1, T2, and T4 significantly increased the germination percentage as compared to the control (T0), yet it was lesser than the germination percentage in treatment T3.

Table 9: Effect of FGDB on germination percentage, plant height, and crop yield during two cropping years

	Germination percentage (%)		Plant height (cm)		Crop yield (kg·hm–2)	
	2011	2012	2011	2012	2011	2012

TO	56.34 ± 1.89c	60.23 ± 1.45c	105.26 ± 1.56c	108.31 ± 1.47c	827.65 ± 32.76c	1050.15 ± 44.82d
T1	62.35 ± 1.09b	69.97 ± 1.78b	119.70 ± 3.41b	124.21 ± 2.96b	1018.22 ± 43.13b	1574.17 ± 53.56b
T2	64.87 ± 2.15b	71.11 ± 1.33b	124.51 ± 2.94b	129.45 ± 2.57b	1113.47 ± 42.10a	1584.47 ± 25.16b
T3	79.02 ± 3.07a	91.25 ± 2.27a	138.52 ± 1.30a	154.52 ± 1.30d	1195.56 ± 36.47a	1678.06 ± 32.28a
T4	77.50 ± 2.89a	88.32 ± 2.56a	135.89 ± 2.24a	143.78 ± 2.14a	1097.36 ± 53.28b	1446.98 ± 40.72c

Letters denote statistically significant differences across treatments (p=0.05).

Plant Height

It is evident from Table 9 that because of the increase in the dosage of FGDB from 4.5 to 13.5 thm^{-2}, the plant height of maize increased significantly ($P \leq 0.05$) and progressively in the range from 13.71 to 29.09% in 2011 and 14.68% to 32.74% over the control (TO). The plant height increase in all the FGDB amended plots was significantly ($P \leq 0.05$) higher than the control. The increase of plant height in treatment T3 was significantly ($P \leq 0.05$) greater than all other FGDB treatments (Table 9). The results indicated that higher FGDB application rates of T4 may retard growth of plant height when compared with application rates of T3. Even at the highest level treatments, however, little negative effects on the plant growth and no symptoms of toxicity were noticed. Hence, the optimum dosage of FGDB is 13.5 thm^{-2} for test site and crop species. The adverse effect of the application of higher dosages of FGDB (beyond 13.5 thm^{-2}) was observed to retard growth of plant height. In general, the addition of FGDB stimulated the increase of plant height.

Crop Yield

Similar to the plant heights, crop yield was also affected by FGDB treatments. In the first year, after harvesting at day 180, the yield of maize was maximum in treatment T3 followed by T2, T4, and T1, respectively, and the increase was significantly more than the control. The changes in crop yield are shown in Table 9. The addition of T1, T2, T3, and T4 resulted in 23.0%, 34.5%, 44.4%, and 32.58% increase in the crop yield in 2011, respectively. In 2012, the crop yield increase at 49.9%, 50.9%, 59.8%, and 37.8%, compared with the control, respectively. The increases in these four treatments were comparable to each other. Similarly, caused by the cumulative effects of FGDB, the yield of the second year has a slight increase comparable to the yields of the first year, respectively. All the four soil amendment treatments significantly improved crop yield when compared with the controls. The improved crop yield possibly was from a decrease in soil pH value and associated changes in soil particle size and other physical properties. The significantly greater average growth parameters and yield in subplots with FGDB application demonstrated that FGDB has the potential to improve crop plant growth and yield in alkaline soils.

DISCUSSION

There are two main aspects to consider when evaluating the efficacy of FGDB as soil amendment materials. One is the ability of the material to neutralize soil alkalinity and lower high soil pH values; the other is the ability to improve soil fertility and enhance the growth and yield of crops.

The agronomic benefit of FGDB applications is primarily associated with improved physicochemical characteristics of the soil. Application of FGDB can reduce clay dispersion and force clay particles into microaggregates; hence, the silt content increased eventually, and better soil texture gradually formed. Bulk density is an indicator of both pore space and solid particles within the profile and provides an indication of water holding capacity and amount of organic matter present. The amendments had some positive effects on the bulk density, because FGDB is generally considered to enhance aggregation, thus reducing bulk density. FGDB with its low bulk density is a suitable additive to improve the bulk density of soils in a mixture. So reductions in bulk density can be attributed to an increase in aggregate content as a result of the FGDB applications. The increase in aggregate content is due to the changes in packing patterns which occur when particles of different size classes are mixed together. FGDB amended soils have less bulk density and greater water holding capacity, compared with non-FGDB amended soil. It means that FGDB has similar function of fly ash, which is reported to have the function of relieving surface crusting and compaction, increasing water infiltration and holding capacity, and improve aggregate stability.

The FGDB addition alters soil physical properties such as its texture, bulk density, water holding capacity [24], and particle size distribution [36]. Campbell et al. (1983) found that FGDB addition at the rate of 10% increased the water holding capacity 7.2 and 413.2 times for fine and coarse sands, respectively [37].

Soil salinity is measured as electrical conductivity (EC). As EC is a quantitative measure of the presence of cations and anions in the soil, FGDB enriched with several cations and anions, on addition to the normal soils, enhanced electrical conductivity of FGDB amended soils significantly.

Due to the better solubility of FGDB by mixing it with soils surface 15 cm soil, was more effective in the removal of exchangeable sodium than FGDB applied on the soil surface. Sodium salts in the alkaline soil are diluted by the application of FGDB that nearly complete displacement of exchangeable sodium by calcium from the FGDB solution occurs. Application of FGDB at different rates decreased the soil pH and ESP, while it increased EC of the alkaline soil at the time of sowing of maize. The amended soils showed a relatively significant decrease in pH and ESP following increased applications of the FGDB. The soil improvement caused by the amendment made conditions more favorable for plant growth.

A correlative relationship between soil pH and the exchangeable sodium percentage for some sodic soils has been pointed out earlier. It means ESP values have a similar change trend to soil pH. The decrease in pH and ESP during reclamation of alkaline soils can be explained due to the replacement of exchangeable Na^+ ions with Ca^{2+}, which is an important mechanism in decreasing pH during reclamation of alkaline soils [38]. For alkali soils, Sakai (2004) found that the amended soils showed a decrease in pH from 10.5 to 9.5 after addition of FGDB [8]. Increase in EC and moisture retaining field capacity were reported in fly ash amended soils for maize [39]. Thus, FGDB also can be considered as an effective liming agent to neutralize solution pH and reduce exchangeable alkalinity values.

The available nutrients in the FGDB amended soil could satisfy the growth demand of plants, and the pH tended to neutrality. Addition of FGDB to soil in the present study has been proved to have some positive effect on nutrients status and other chemical parameters. According to M. R. Khan and M. W. Khan, a gradual increase in FGDB concentrations in the normal field soil improved the availability of sulfate, carbonate, bicarbonate, chloride, P, K, Ca, Mg, Mn, Cu, Zn, and B [40]. FGDB has successfully been used to quickly reduce Na^+ concentrations during the reclamation of sodic soils. The results of the present study confirm the findings of earlier works showing decrease in Na^+ concentrations with increase of FGDB incorporation [41]. Better growth and yield may be due to the presence of some utilizable plant nutrients such as sulphate, P, K, and Ca in FGDB because field soil used in the experiment was low on plant nutrients especially P and K.

There is no clear guidance we can follow with regard to the application rate of FGDB as soil amendment. The FGDB significantly increased some important plant macronutrients; the possibility exists that, at a rational rate of application, crop yield increase can be attained in FGDB amended soils. FGDB in heavy dose may have some potential adverse effect on plant growth. Addition of FGDB to soil up to 13.5 t/hm^2 was beneficial for growth and yield of maize and the physical and chemical properties of test soil, above which it has an adverse effect on both growth and yield of maize. The problem may not limit the potential use of FGDB with crops. However, more emphasis should be paid to the optimum application rate of FGDB with certain soil and certain crop species and how well the crop can withstand the potential deleterious effects.

The FGDB was beneficial to plant growth if applied to alkaline soil at appropriate levels. In the present study, treatments having FGDB as a soil amendment demonstrated that FGDB applications had a positive impact on average yield and germination percentage and plant height. The results indicated that the growth parameters like germination percentage and plant height increased with the four different FGDB application rates. The optimum dosage of FGDB is 13.5 t/hm^2 to the test crop, maize. Treatment with above 13.5 t hm^{-2} FGDB as soil amendment showed an adverse effect on the plant growth by retarding the plant height. Mittra et al. (2005) observed better growth of rice plants grown on soil amended with FGDB at lower concentrations [42]. Singh et al. (1997) found that higher FGDB application either delays or drastically inhibits plant growth, development, and other specific parameters. The retarded growth of plants grown with higher concentrations of FGDB amended soil was reflected by slow growth as well as grain yield [43].

According to the previous experiments, both in field and greenhouse conditions showed that different rates of FGDB can promote plant growth and improve the agronomic properties of the soils [13, 16]. Due to the improved agronomic soil properties, crop usually grows more readily in FGDB amended soils. Enhanced growth with application of FGDB has been reported in many plants [17]. Growth of a new planting of alfalfa (Medicago sativa L.) was increased by 10 to 40% by the treatments compared to the untreated control [44]. As an industrial waste, the availability of FGDB can be a supply of nonconventional sources of plant nutrients and soil conditioners and can be utilized for increasing crop yield and can help to improve the environment and ecology as the waste can be utilized for enhancing yield and will also help to solve the disposal problem of wastes. In China, use of these large quantities of FGDB in agricultural land not only improves the physical and chemical fertility of soils but also resolves industrial waste disposal problem.

CONCLUSIONS

Application of FGDB as an alkaline soil amendment positively boosted the soil quality by improving its physical and chemical properties. Based on the results, compared to zero application of FGDB, treatments modified significantly the soil physical (soil texture, bulk density, and water holding capacity) and chemical (soil pH and EC and plant nutrients) properties at field scale. These modifications finally clearly affected some selected soil physiochemical properties and crop growth parameters including germination percentage, plant height, and crop yield of maize. Of all 4 different treatments, we obtained the most positive effect at the rate of 13.5 t hm^{-2} treatment T3. The physical benefits of the soil include improved soil texture due to the addition of silt-sized FGDB that promoted better soil mechanical composition, water holding capacity, and bulk density. Chemical benefits may be due to the presence of essential plant nutrients for crop production or the supply of a more favorable substrate for plant growth by modifying the soil. The liming capacity of the FGDB makes it suitable for decreasing pH in alkaline soils. FGDB also has beneficial effects on maize growth which is attributed to its ability to supply essential plant nutrients. Its addition to soil resulted in agronomic benefits enabling raising germination percentage, plant height, and crop yield and even growing of crops on nutrient deficient soils and can be successfully used to correct the soil pH values.

Our results confirmed that FGDB can be successfully used to buffer alkaline soils and adjust soil pH to a suitable range that optimizes the availability of most of the plant nutrients. However, in future studies, a long-term experiment should be carried out to assess the ecological safety effects of agricultural use of FGDB even at low doses.

ACKNOWLEDGMENTS

This study was financially supported by the open program from the State Key Laboratory of Earth Surface Process and Resource Ecology (Grant no. 2010-KF-07) and by the National Natural Science Foundation of China Program (Grant no. 41261068). The authors are thankful to Ms. Linhong Wu for providing valuable help in collecting and analyzing soil samples.

REFERENCES

1. G. I. Metternicht and J. A. Zinck, "Remote sensing of soil salinity: potentials and constraints," Remote Sensing of Environment, vol. 85, no. 1, pp. 1–20, 2003.
2. M. P. Reddy, M. T. Shah, and J. S. Patolia, "Salvadora persica, a potential species for industrial oil production in semiarid saline and alkali soils," Industrial Crops and Products, vol. 28, no. 3, pp. 273–278, 2008.

3. J. Y. Li, J. He, and J. Wang, "Safety evaluation index system of blackish water irrigation," Journal of China Rural Water and Hydropower, no. 8, pp. 8–12, 2010 (Chinese).
4. Y. Sakai, S. Shitara, and M. Sadakata, "Salts transport in alkali soil reclamation by gypsum and prediction of Na leaching in field in China," Journal of Water and Environment Technology, vol. 7, no. 2, pp. 121–133, 2009.
5. American Coal Ash Association, "The changing face of the utility industry and its impact on coal combustion by-product management," in Proceedings of the 12th International Symposium on Management and Use of Coal Combustion by-Products, pp. 26–30, American Coal Ash Association, Orlando, Fla, USA, January 1997.
6. R. K. Srivastava, W. Jozewicz, and C. Singer, "SO_2 scrubbing technologies: a review," Environmental Progress, vol. 20, no. 4, pp. 219–227, 2001.
7. M. Ahmaruzzaman, "A review on the utilization of fly ash," Progress in Energy and Combustion Science, vol. 36, no. 3, pp. 327–363, 2010.
8. Y. Sakai, S. Matsumoto, and M. Sadakata, "Alkali soil reclamation with flue gas desulfurization gypsum in China and assessment of metal content in corn grains," Soil and Sediment Contamination, vol. 13, no. 1, pp. 65–80, 2004.
9. F. E. Rhoton and D. S. McChesney, "Influence of FGD gypsum on the properties of a highly erodible soil under conservation tillage," Communications in Soil Science and Plant Analysis, vol. 42, no. 16, pp. 2012–2023, 2011.
10. L. Chen, R. Stehouwer, M. Wu et al., "Minesoil response to reclamation by using a flue gas desulfurization product," Soil Science Society of America Journal, vol. 77, no. 5, pp. 1744–1754, 2013.
11. M. E. Buckley and R. P. Wolkowski, "In-season effect of flue gas desulfurization gypsum on soil physical properties," Journal of Environmental Quality, vol. 43, no. 1, pp. 322–327, 2014.
12. T. M. DeSutter, L. J. Cihacek, and S. Rahman, "Application of flue gas desulfurization gypsum and its impact on wheat grain and soil chemistry," Journal of Environmental Quality, vol. 43, no. 1, pp. 303–311, 2014.

13. S. Dwivedi, R. D. Tripathi, S. Srivastava et al., "Growth performance and biochemical responses of three rice (Oryza sativa L.) cultivars grown in fly-ash amended soil," Chemosphere, vol. 67, no. 1, pp. 140–151, 2007.
14. A. Singh, R. K. Sharma, and S. B. Agrawal, "Effects of fly ash incorporation on heavy metal accumulation, growth and yield responses of Beta vulgaris plants," Bioresource Technology, vol. 99, no. 15, pp. 7200–7207, 2008.
15. V. C. Pandey and N. Singh, "Impact of fly ash incorporation in soil systems," Agriculture, Ecosystems & Environmen, vol. 136, no. 1-2, pp. 16–27, 2010.
16. A. Singh and S. B. Agrawal, "Response of mung bean cultivars to fly ash: growth and yield,"Ecotoxicology and Environmental Safety, vol. 73, no. 8, pp. 1950–1958, 2010.
17. S. Jala and D. Goyal, "Fly ash as a soil ameliorant for improving crop production—a review," Bioresource Technology, vol. 97, no. 9, pp. 1136–1146, 2006.
18. K. Dontsova, Y. B. Lee, B. K. Slater, and J. M. Bigham, Gypsum for Agricultural Use in Ohio: Sources and Quality of Available Products, Ohio State University School of Natural Resources, 2005.
19. D. M. Endale, H. H. Schomberg, D. S. Fisher, D. H. Franklin, and M. B. Jenkins, "Flue gas desulfurization gypsum: implication for runoff and nutrient losses associated with broiler litter use on pastures on ultisols," Journal of Environmental Quality, vol. 43, no. 1, pp. 281–289, 2014.
20. H. Lee, H. S. Ha, C. H. Lee, Y. B. Lee, and P. J. Kim, "Fly ash effect on improving soil properties and rice productivity in Korean paddy soils," Bioresource Technology, vol. 97, no. 13, pp. 1490–1497, 2006.
21. G. L. Terman, V. J. Kilmer, C. M. Hunt, and W. Buchanan, "Fluidized bed boiler waste as a source of nutrients and lime," Journal of Environmental Quality, vol. 7, no. 1, pp. 147–150, 1978.
22. J. T. Crews and W. A. Dick, "Liming acid forest soils with flue gas desulfurization by-product: growth of Northern red oak and leachate water quality," Environmental Pollution, vol. 103, no. 1, pp. 55–61, 1998.

23. M. E. Schutter and J. J. Fuhrmann, "Soil microbial community responses to fly ash amendment as revealed by analyses of whole soils and bacterial isolates," Soil Biology and Biochemistry, vol. 33, no. 14, pp. 1947–1958, 2001.
24. A. C. Chang, L. J. Lund, A. L. Page, and J. E. Warneke, "Physical properties of fly ash amended soils,"Journal of Environmental Quality, vol. 6, no. 3, pp. 267–270, 1977.
25. R. B. Clark, K. D. Ritchey, and V. C. Baligar, "Benefits and constraints for use of FGD products on agricultural land," Fuel, vol. 80, no. 6, pp. 821–828, 2001.
26. Y. Zhang, E. Kendy, Y. Qiang, L. Changming, S. Yanjun, and S. Hongyong, "Effect of soil water deficit on evapotranspiration, crop yield, and water use efficiency in the North China Plain," Agricultural Water Management, vol. 64, no. 2, pp. 107–122, 2004.
27. Y. B. Lin, Experimental study on the amelioration of coastal salt-affected soils in north China based on winter brackish ice mulching [Dissertation], Beijing Normal University, Beijing, China, 2011.
28. A. L. Black and J. F. Power, "Effect of chemical and mechanical fallow methods on moisture storage, wheat yields, and soil erodibility," Soil Science Society of America, vol. 29, no. 4, pp. 465–468, 1965.
29. M. Guo, N. Tongtavee, and M. Labreveux, "Nutrient dynamics of field-weathered delmarva poultry litter: implications for land application," Biology and Fertility of Soils, vol. 45, no. 8, pp. 829–838, 2009·
30. U.S. EPA, Method 3050. Acid Digestion of Sediments, Sludges, Soils and Oils. SW-846, U.S. EPA, Washington, DC, USA, 1996.
31. N. Koukouzas, C. Ketikidis, and G. Itskos, "Heavy metal characterization of CFB-derived coal fly ash,"Fuel Processing Technology, vol. 92, no. 3, pp. 441–446, 2011.
32. B. Wang, G. J. Xiao, G. L. Mao, Z. H. Yue, and X. Xu, "coal-fired flue gas desulfurated waste residue application on salinealkali soil amelioration and oil-sunflower growth," Chinese Journal of Plant Ecology, vol. 34, no. 10, pp. 1227–1235, 2010, (in Chinese with English abstract).

33. U.S. EPA, "Standards for the use or disposal of sewage sludge: final rules," Federal Register, vol. 58, 40 CFR Parts 503, Subpart D, pp. 9398–9400, 1993.
34. P. Asokan, M. Saxena, and S. R. Asolekar, "Coal combustion residues—environmental implications and recycling potentials," Resources, Conservation and Recycling, vol. 43, no. 3, pp. 239–262, 2005.
35. Council on Soil Testing and Plant Analysis, Handbook on Reference Methods for Soil Analysis, Soil and Plant Analysis Council, Athens, Ga, USA, 1992.
36. S. Sharma, M. H. Fulekar, and C. P. Jayalakshmi, "Fly ash dynamics in soil-water systems," Critical Reviews in Environmental Control, vol. 19, no. 3, pp. 251–275, 1989.
37. D. J. Campbell, W. E. Fox, R. L. Aitken, and L. C. Bell, "Physical characteristics of sands and amended with fly-ash," Australian Journal of Soil Research, vol. 21, no. 2, pp. 147–154, 1983.
38. S. Gaind and A. C. Gaur, "Quality assessment of compost prepared from fly ash and crop residue,"Bioresource Technology, vol. 87, no. 1, pp. 125–127, 2003.
39. N. Kalra, M. C. Jain, H. C. Joshi, et al., "Soil properties and crop productivity as influenced by flyash incorporation in soil," Environmental Monitoring and Assessment, vol. 87, no. 1, pp. 93–109, 2003.
40. M. R. Khan and M. W. Khan, "The effect of fly ash on plant growth and yield of tomato," Environmental Pollution, vol. 92, no. 2, pp. 105–111, 1996.
41. E. Amezketa, R. Aragüés, and R. Gazol, "Efficiency of sulfuric acid, mined gypsum, and two gypsum by-products in soil crusting prevention and sodic soil reclamation," Agronomy Journal, vol. 97, no. 3, pp. 983–989, 2005.
42. B. N. Mittra, S. Karmakar, D. K. Swain, and B. C. Ghosh, "Fly ash—a potential source of soil amendment and a component of integrated plant nutrient supply system," Fuel, vol. 84, no. 11, pp. 1447–1451, 2005.
43. S. N. Singh, K. Kulshreshtha, and K. J. Ahmad, "Impact of fly ash soil amendment on seed germination, seedling growth and metal composition of Vicia faba L," Ecological Engineering, vol. 9, no. 3-4, pp. 203–208, 1997.

44. L. Chen, W. A. Dick, and S. Nelson Jr., "Flue Gas Desulfurization products as sulfur sources for alfalfa and soybean," Agronomy Journal, vol. 97, no. 1, pp. 265–271, 2005.

Chapter 4

Ectomycorrhizal Influence on Particle Size, Surface Structure, Mineral Crystallinity, Functional Groups, and Elemental Composition of Soil Colloids from Different Soil Origins

Yanhong Li, Huimei Wang, Wenjie Wang, Lei Yang, and Yuangang Zu

The Key Laboratory of Forest Plant Ecology Ministry of Education, Harbin, Heilongjiang 150040, China

ABSTRACT

Limited data are available on the ectomycorrhizae-induced changes in surface structure and composition of soil colloids, the most active portion in soil matrix, although such data may benefit the understanding of mycorrhizal-aided soil improvements. By using ectomycorrhizae (Gomphidius viscidus) and soil colloids from dark brown forest soil (a good loam) and saline-alkali soil (heavily degraded soil), we tried to approach the changes here. For the good loam either from the surface or deep soils, the fungus treatment induced physical absorption of covering materials on colloid surface with nonsignificant increases in soil particle size ($P > 0.05$). These increased the amount of variable functional groups (O–H stretching and bending, C–H stretching, C=O stretching, etc.) by 3–26% and the crystallinity of variable soil minerals (kaolinite, hydromica, and quartz) by 40–300%. However, the fungus treatment of saline-alkali soil obviously differed from the dark brown forest soil. There were 12–35% decreases in most functional groups, 15–55% decreases in crystallinity of most soil minerals but general increases in their grain size, and significant increases in soil particle size ($P<0.05$). These different responses sharply decreased element ratios (C : O, C : N, and C : Si) in soil colloids from saline-alkali soil, moving them close to those of the good loam of dark brown forest soil.

INTRODUCTION

At the global scale, soil degradation, including soil erosion, is a potential threat to food security, and phytorehabilitation measures for controlling soil degradation are a popular and urgent topic of research [1,2]. Ectomycorrhizal (ECM) fungi and associated symbionts can promote the growth of plants and increase their tolerance to unfavorable soil conditions such as nutrient deficiency or heavy metal pollution [3–5]. More than 5000 fungi can form ECM symbionts with over 2000 woody plants [6], showing the importance of ectomycorrhizae in plant-soil interactions. Commercially available mycorrhizal inocula, which consist of a single fungus species, are currently used for afforestation and grassland recovery [7]. ECM fungi enhance the growth and fitness of plants [8–11] by providing them with mineral and organic nutrients from the soil matrix and by protecting the carbohydrates and organic compounds that are stored in the roots from pathogenic organisms [12]. Improvements in root length, soil P utilization efficiency, and disease and stress resistance, as well as enhanced soil nutrient availability, have been reported [13, 14]. Recent studies have been performed on the identification of ECM fungi [15–18], interactions among various fungi and their effects on soil pollution rehabilitation [17, 19], and the underlying genetic basis for ECM functions [20–22]. These studies have provided a sound basis for understanding the mechanisms of the interaction between ECM fungi and various plants. Mycorrhizal fungi can also directly stabilize soil both through their hyphal network and through the secretion of glue-like chemicals [7]. However, the rarity of studies on the interaction between soil particles and ECM fungi hinders a full understanding of the function of fungi in soil health maintenance and soil physical texture formation.

Owing to the profound heterogeneity among soil samples from field sampling campaigns, it is difficult to visualize the interaction between ECM fungi and soil aggregates at the microscale of millimeters or nanometers. Soil colloids are generally considered to be particles with effective diameters of around 10 nm to 10 μm, with the smallest colloids just larger than dissolved macromolecules and the largest colloids being those that resist settling once suspended in soil pore water [23–25]. Soil colloids are the most active portion of the soil and largely determine the physical and chemical properties of the soil. Organic colloids are more reactive chemically and generally have a greater influence on soil properties per unit weight than inorganic colloids. Inorganic colloids of clay minerals are usually crystalline (although some are amorphous) and usually have a characteristic chemical and physical configuration. These features of soil colloids made it more suitable to study the interaction between ECM fungi and soil particles in indoor laboratory tests compared to bulk soils with heterogeneous composition [25]. In this paper, one aim was to find conformational changes in fungal extracts induced by the interaction with soil colloidal particles.

Previous papers have described good methods for identifying the surface structure and composition of soil colloids. To examine surface changes and microstructure changes in soil colloids after the addition of soil conditioner, Wang et al. used atomic force microscopy (AFM) to characterize 3D structural changes and used scanning electron microscopy (SEM) to find 2D surficial changes [26]. Laser particle-size analyzers have been used in a variety of studies with soil samples [27–30]. X-ray powder diffraction (XRD) is a nondestructive and rapid analytical technique primarily used for phase identification of soil minerals that can provide information regarding grain size and relative crystallinity, for example, interactions between clay minerals and organic matter in relation to carbon sequestration [31, 32]. Infrared spectroscopy is a well-established technique for the identification of chemical compounds and specific functional groups in compounds and, thus, is a useful tool for soil applications [33–35]. X-ray photoelectron spectroscopy (XPS) has the advantage of being able to detect all elements in the soil (except for H and He) and provides much valuable information on the composition of, and bonding state of elements in, surface and near-surface layers of many minerals [36–38]. The above methods may be useful for clarifying the influence of ECM fungus extracts on the surface structure and composition of soil colloids.

To reveal the underlying mechanism of the impact of ECM fungi on soil particles, soil colloids were extracted from 3 types of soils, including dark brown forest soil (deep and surface layers) and saline-alkaline soil in a grassland (surface layer). After fungal extract treatment, the laser particle-size analyzer, AFM, SEM, XRD, IR, and XPS were used to characterize the structural and surface changes. We hypothesized that ECM fungi would have different effects on soil colloids from different origins, and this could contribute to the improvement of degraded soil and the formation of healthy soil structure. The main aim of this study was to investigate the relationship between ECM fungi and soil and reveal the mechanism underlying soil improvement.

MATERIAL AND METHODS

Preparation of Soil Colloids

Soil samples were collected from the top soil (0–20 cm) in a typical saline-alkali region of the Songnen Plain (45°59′55″N, 124°29′48″E). As a nonsaline loam control, we used dark brown soil from the surface layer (0–20 cm) and a deeper layer (60–80 cm) from the Experimental Forest Farm of Northeast Forestry University (45°43′6″N, 126°37′54″E).

The soil colloids were separated according to [25, 26] as follows. One gram of air-dried soil was fully dispersed in 100 mL of ultrapure water in a 250 mL beaker. The suspension was allowed to stand undisturbed for more than 24 h. Sands and silts in the soil sample were gradually deposited at the bottom of the beaker, whereas the soil colloids were left in the suspension indicated as a turbid solution. The upper suspension was carefully decanted to a centrifuge tube and then centrifuged at 12000 rpm for 10 min. The precipitates were dissolved with 25 mL of ultrapure water and regarded as the soil colloidal solution.

Preparation of Fungus Extract

Gomphidius viscidus was sampled from Inner Mongolia, China, for laboratory testing. It is an important mycorrhizal fungus in coniferous forests and belongs to the Gomphidiaceae family of the Agaricales [40]. It was grown in modified Melin-Norkrans (MMN) medium: $CaCl_2$, 0.05 g; NaCl, 0.025 g; KH_2PO_4, 0.5 g; $(NH_4)_2HPO_4$, 0.25 g; $MgSO_4 \cdot 7H_2O$, 0.15 g; $FeCl_3$ (1%), 1.2 mL; thiamine HCl, 0.2 mL; malt extract, 3 g; glucose, 10 g; stock solution of micronutrients (contents per liter: H_3BO_3, 2.86 g; $MnCl_2$, 1.81 g; $ZnSO_4$, 0.22 g; $CuSO_4$, 0.08 g; and $NaMoO_4$, 0.02 g), 1 mL; ultrapure water, 1000 mL; and 15 g of agar in the case of agar media. The pH of the media was adjusted to 5.45–5.55 before autoclaving (121°C, 0.1 Mpa, and 20 min). The medium was dispersed aseptically in a 10 cm culture dish and stored in a 4°C refrigerator. The strain was inoculated on the MMN solid culture medium, training repeatedly until no other bacteria were produced [41].

Next, liquid medium was used, and 200 mL aliquots of the medium were dispensed into 250 mL beakers and autoclaved at 117°C for 20 min. Each Erlenmeyer flask was inoculated with fungal colonies of G. viscidus and then cultured on a shaking table. After 14 days, the fungal hyphae were used to extract fungal solution. First, culture media were filtered using a 4-layer gauze. After being cleaned repeatedly with ultrapure water, the mycelium was ground in a mortar and centrifuged (12000 rpm) at 4°C for 10 min, and the supernatant was collected as the fungus extract solution.

Particle Size and Surface Structure Observation

After the addition of fungus extract solution to soil colloids from different origins, the particle size and surface structure were determined with a laser particle-size analyzer (ZetaPALS, USA), an atomic force microscope (AZ, USA), and a scanning electron microscope (Quanta 200, FEI, USA). Methods were revised from [26, 30], and detail was as follows.

For the laser size analysis, 10 μL of fungus extract was added to 4 mL of soil colloid solution (about 1 mg soil colloids), and the fungus extract was replaced by 10 μL of ultrapure water in the control group. Twelve hours after mixing, the mixture was analyzed using the laser particle-size analyzer. For AFM observation, 10 μL of fungus extract was added to 4 mL of soil colloid solution, and the fungus extract was replaced by 10 μL of ultrapure water in the control group. One hour after mixing, centrifugation (12000 rpm) at 25°C for 10 min was used to precipitate the soil colloids. This colloid precipitate was then dissolved with 4 mL of ultrapure water, dropped on the mica surface, and air-dried before AFM imaging. The AFM images of the soil colloids were obtained in tapping mode by using the PicoPlus II AFM system from Molecular Imaging (MI) Corporation (AZ, USA). For SEM observation, the same procedures as those for AFM observation were used, and the soil colloidal solution was dropped on the sample stage and then air-dried before SEM imaging (Quanta 200, FEI, USA). The samples were sputter-coated with a thin layer of gold-palladium (5–10 nm, 25 mA, and 3 min) at room temperature by using a sputter coater before the examination.

Crystal Structure Diffraction of Soil Minerals

The preparation of treatments and controls was the same as that for AFM, and SEM and X-ray powder diffraction (XPD) observation were carried out according to [32]. After being air-dried, X-ray powder diffraction patterns were collected in transmission by using an X-ray powder diffractometer (D/Max 2200, Rigaku, Japan) with a rotating anode (Philips) and Cu $K\alpha_1$ radiation generated at 30 mA and 40 kV. The range of 2θ diffraction angles examined was 10–40° with steps of 0.02° and a measuring time of 0.3 s per step.

Observations of Functional Group Change

The preparation of treatments and controls was the same as that for AFM, and SEM and infrared spectrum (IR) measurements were carried out according to [33–35]. The samples were diluted with 1% KBr mixing powder and separately pressed to obtain self-supporting disks. Tablets for IR measurements were prepared by pressing the powder mixture at a load of 8 tons for 8 min. The IR spectrum was obtained by a compact Fourier transform infrared spectrophotometer (IR Affinity-1, SHIMADZU, Japan) and recorded across a wave number range of 4000–500 cm^{-1} at a resolution of 4 cm^{-1}.

Analysis of the Atomic Concentration of Elements

A K-Alpha spectrometer equipped with a concentric hemispherical analyzer in the standard configuration (Thermo Scientific, USA) was used in this analysis with the method revised from [38]. After the preparation of control and treatment samples (the same as that for SEM and AFM), the soil colloidal solution was dropped on a cleaned HOPG surface and then air-dried before examination. The vacuum system consisted of a turbomolecular pump and a titanium sublimation pump. The residual pressure before the analysis was lower than 10^{-7} Pa. The X-ray source was AlKα, and it was run at 30 mA and 80 kV. The incident angle was 49.1°, and the emission angle was 0° with respect to the sample's surface normal. All the spectra were obtained in digital mode. The wide-scan spectra were acquired from 1000 eV to 0 eV. Sample charging was corrected by comparing all binding energies to the adventitious carbon at 285.0 eV. Detailed spectra were processed using CasaXPS software (V2.3.12, Casa Software Ltd., UK). An iterated Shirley-Sherwood background subtraction was applied before peak fitting using a nonlinear least-squares algorithm. The atomic concentration of elements was calculated using the software.

Data Analysis

In the analysis of XRD data, the original data were rectified using the Jade program to eliminate Kα and then obtain the XRD pattern for a sample. The upper area (α_c), which was separated with the smooth curve connecting each point of minimum intensity, corresponded to the crystalline portion, and the lower area was the background containing the amorphous portion (α_b). The Jade program was used to calculate grain size and relative crystallinity (relative crystallinity = $\alpha_c / (\alpha_c + \alpha_b)$) [42] and check the effect of the addition of fungus extracts. The major components of the soil were obtained in conjunction with the database and according to the books written by Xie [43] and Marc and Jacques [44].

Six functional groups [45] were selected for IR analysis, as shown in Figure 5: 3750–3300 cm^{-1} is O–H stretching of structural OH; 1200–970 cm^{-1} is Si–O–Si stretching; 950–820 cm^{-1} is O–H bending of structural OH; 2970–2820 cm^{-1} is aliphatic C–H stretching; 1750–1630 cm^{-1} is C=O stretching of carboxylic acids, amides, and ketones; and 1650–1360 cm^{-1} is carbonates.

RESULTS

Results Obtained Using the Laser Particle-Size Analyzer

The particle size and distribution of soil colloids from dark brown forest soil (surface and deep layers) and saline-alkali soil with and without the addition of fungus extracts are shown in Figure 1. After the addition of the fungus extract, the particle size of soil colloids from dark brown forest soil in the surface layer was larger than that in the control, and similar results were found for deep soils. However, the differences were not significant between the treatment and the control ($P>0.05$). In contrast, the addition of the fungus extract increased the particle size of colloids from saline-alkaline soil from 472 ± 11.3 nm to 502 ± 4.0 nm, and this 6.29% increase was statistically significant ($P<0.05$) (Figure 1).

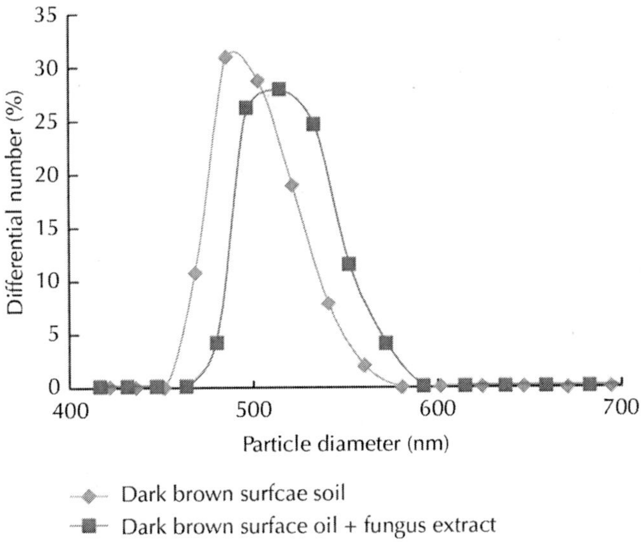
35
30
25
20
15
10
5
0
Differential number (%)
400
500
600
700
Particle diameter (nm)
Dark brown surfcae soil
Dark brown surface oil + fungus extract

(a)

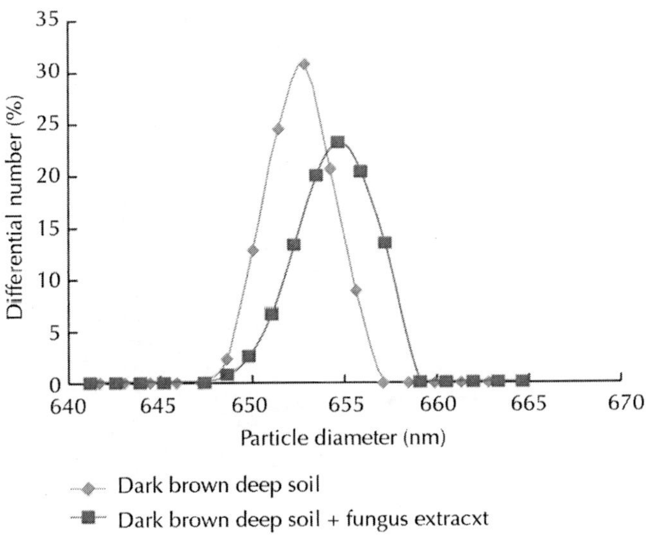
35
30
25
20
15
10
5
0
Differential number (%)
640
645
650
655
660
665
670
Particle diameter (nm)
Dark brown deep soil
Dark brown deep soil + fungus extracxt

(b)

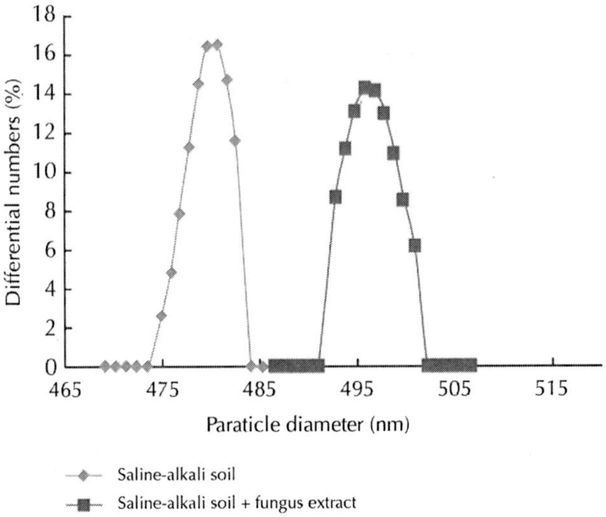

(c)

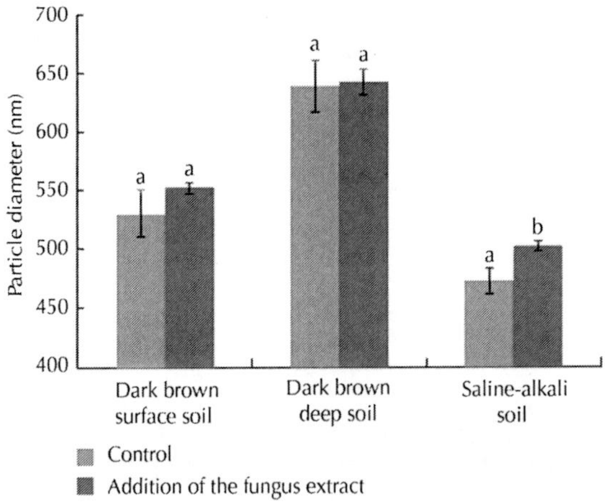

(d)

Figure 1: Particle diameter changes in soil colloids with and without fungus extract addition.

AFM Results

As shown in Figure 2, soil colloids from the surface and deep layers of dark brown forest soil showed similar tendencies. More soil colloids aggregated together and became larger than those in the control. Moreover, more covering materials were found on the surface of these colloids than those in the control samples. Contrary to the soil colloids from the dark brown forest soil, soil colloids from the saline-alkali soil were dispersed after the addition of fungus extracts, and the interaction between the different particles was not as dense as that in the control (Figure 2).

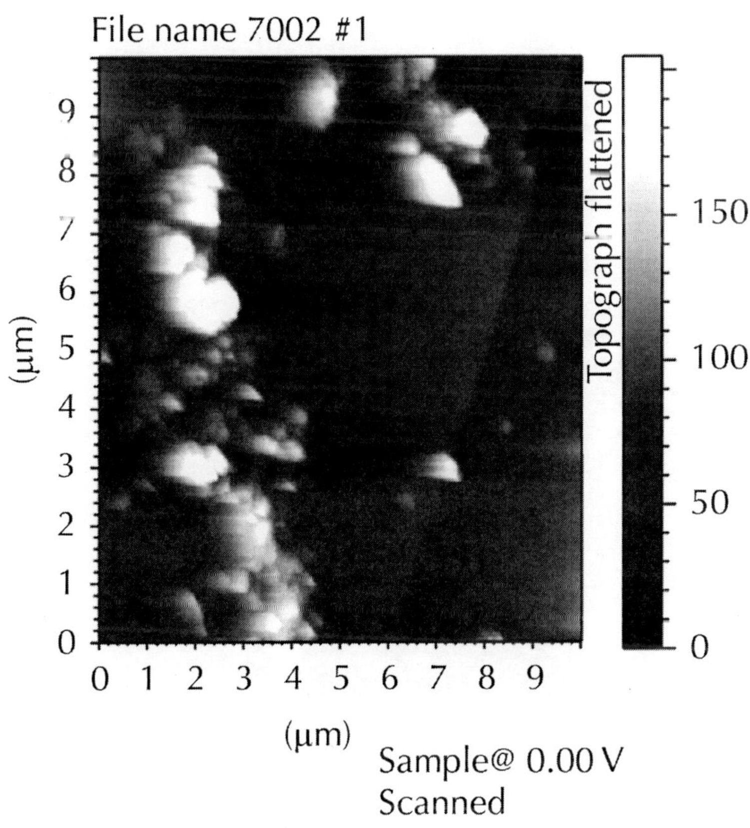

(a)

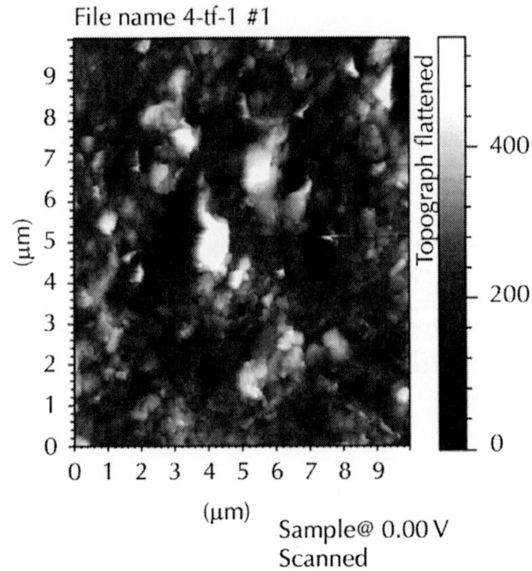
File name 4-tf-1 #1
Topograph flattened
400
200
0
9
8
7
6
5
4
3
2
1
0
(µm)
0 1 2 3 4 5 6 7 8 9
(µm)
Sample@ 0.00 V
Scanned

(b)

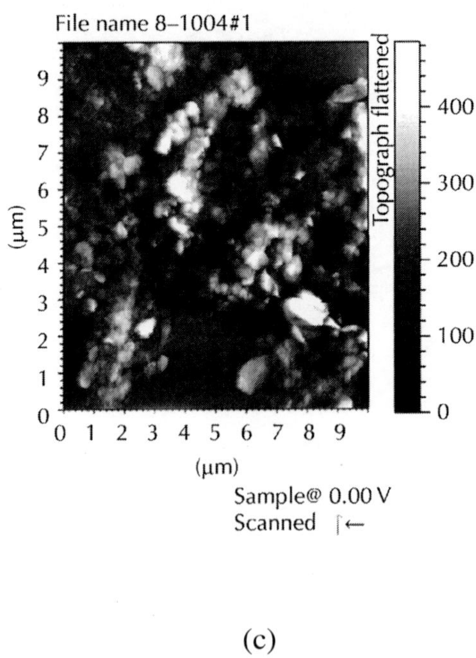
File name 8–1004#1
Topograph flattened
400
300
200
100
0
9
8
7
6
5
4
3
2
1
0
(µm)
0 1 2 3 4 5 6 7 8 9
(µm)
Sample@ 0.00 V
Scanned

(c)

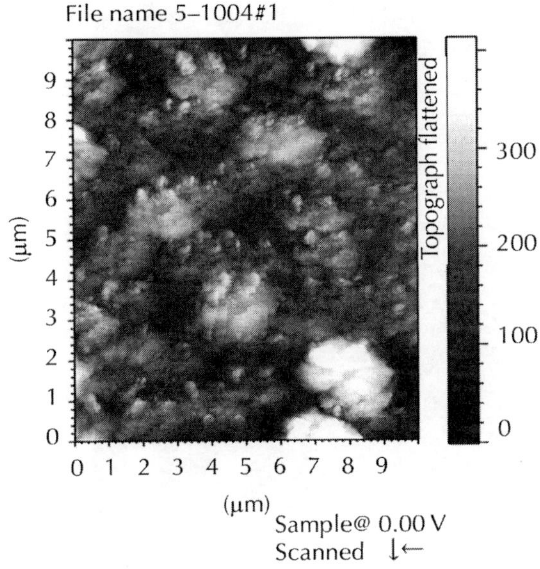
File name 5–1004#1
9
8
7
6
5
4
3
2
1
0
(μm)
0 1 2 3 4 5 6 7 8 9
(μm)
Topograph flattened
300
200
100
0
Sample@ 0.00 V
Scanned ↓←

(d)

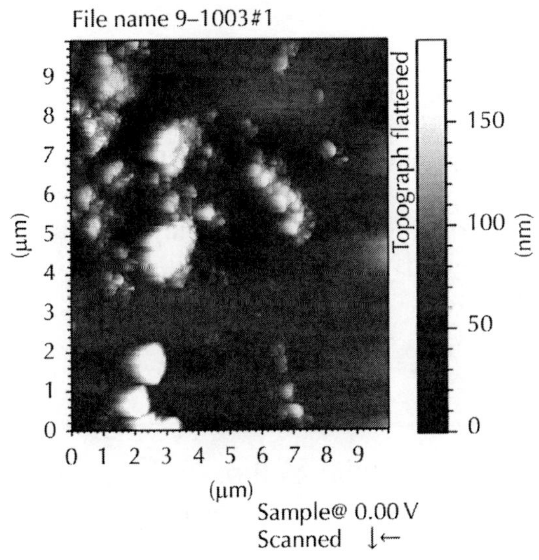
File name 9–1003#1
9
8
7
6
5
4
3
2
1
0
(μm)
0 1 2 3 4 5 6 7 8 9
(μm)
Topograph flattened
150
100
50
0
(nm)
Sample@ 0.00 V
Scanned ↓←

(e)

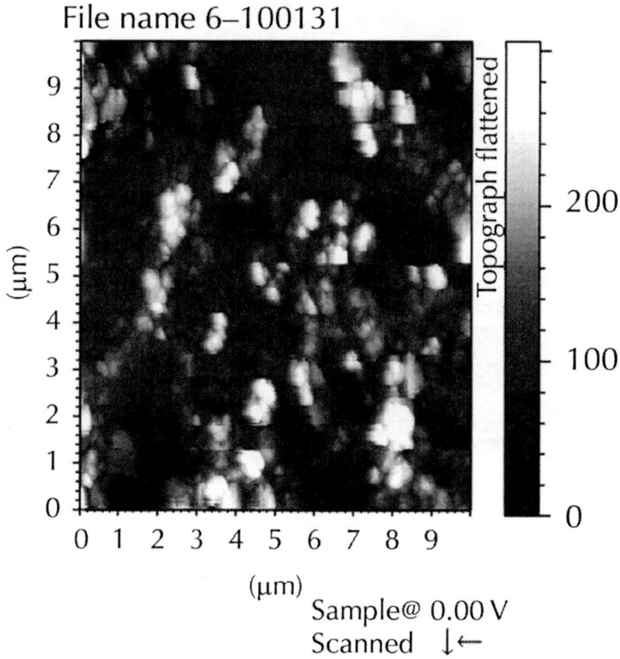

(f)

Figure 2: Atomic force microscopy images of soil colloids with (b, d, and f) and without (a, c, and e) fungus extract addition. (a) Colloids from the surface layer of dark brown forest soil; (b) colloids from the surface layer of dark brown forest soil + fungus extract; (c) colloids from the deep layer of dark brown forest soil; (d) colloids from the deep layer of dark brown forest soil + fungus extract; (e) colloids from saline-alkali soil; (f) colloids from saline-alkali soil + fungus extract.

SEM Results

The results of SEM were similar to those of AFM (Figure 3). After the addition of the fungus extract, the edges of colloid particles from the dark brown forest soil became much smoother, the small gap between soil colloid particles became invisible, and some covering materials seemed to fill these gaps and made the edges less distinct. This tendency was more evident in surface soils than in deep soils (Figures 3(a)–3(d)).

4/6/2011 4:26:23 PM
Mag 10000x
Spot 3.5
WD 10.3 mm
HV 15.0 kV
Pressure
Det ETD
2.0µm
test

(a)

4/6/2011
Mag
Spot 3.5
WD 10.8 mm
HV 15.0 kV
Pressure
Det ETD
2.0µm
test

(b)

(c)

4/6/2011
5:18:01 PM
Mag
10000x
Spot
3.5
WD
10.6 mm
HV
5.0 kV
Pressure
Det
LTD
2.0μm
test

(d)

(c)

(f)

Figure 3: Scanning electron microscopy images of soil colloids with (b, d, and f) and without (a, c, and e) fungus extract addition. The labels are the same as those for Figure 2.

However, results from the saline-alkali soil were different from those from the dark brown forest soil; relatively larger particles with more distinct and acute edges were observed after the addition of the fungus extract (Figures 3(e) and 3(f)).

XRD Results

Both the surface and deep layers of dark brown forest soil had 4 obvious diffraction peaks located at 12.3°, 17.8°, 25.0°, and 26.7°. These peaks are indicative of 3 kinds of minerals: kaolinite (7.15 Å, 3.56 Å), hydromica (4.98 Å), and quartz (3.34 Å) (Figures 4(a) and 4(b)). Soil colloids from saline-alkali soil had 6 diffraction peaks, located at 12.3°, 17.8°, 18.6°, 25.0°, 26.7°, and 29.3° indicating that the main soil mineral composition was kaolinite (7.15 Å, 3.56 Å), hydromica (4.98 Å), vermiculite (4.74 Å), quartz (3.34 Å), and calcite (3.03 Å) [44] (Figure 4(c)).

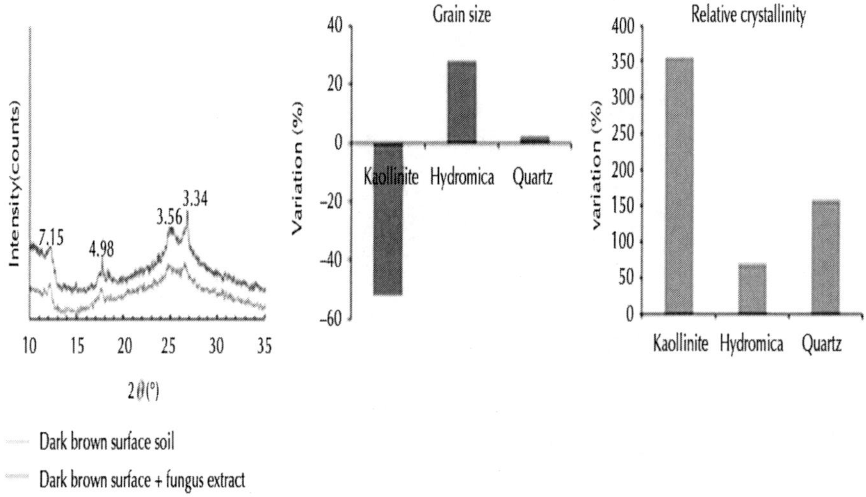

(a)

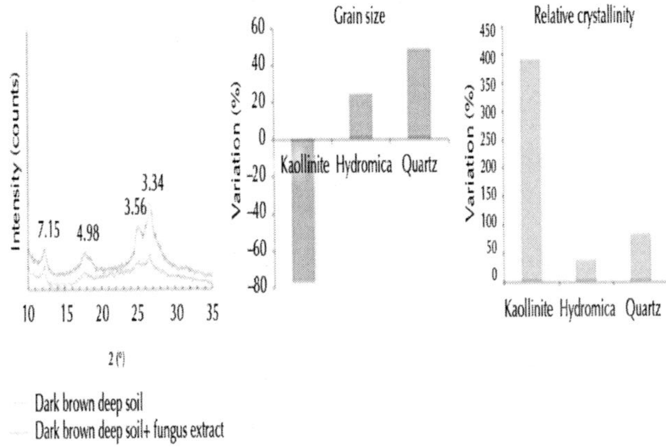

(b)

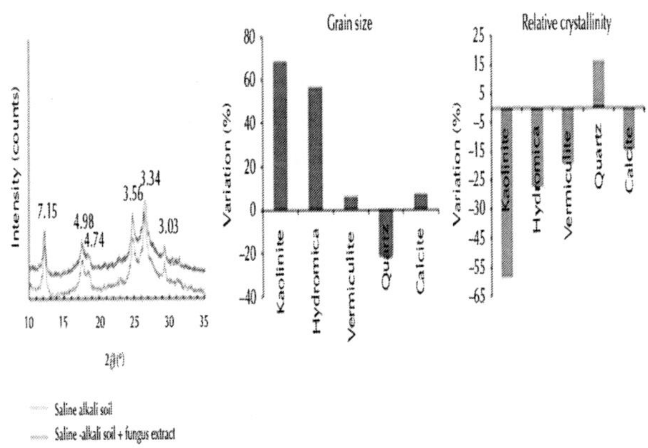

(c)

Figure 4: X-ray powder diffraction results with and without fungus extract addition. (a) Colloids from the surface layer of dark brown forest soil; (b) colloids from the deep layer of dark brown forest soil; (c) colloids from saline-alkali soil.

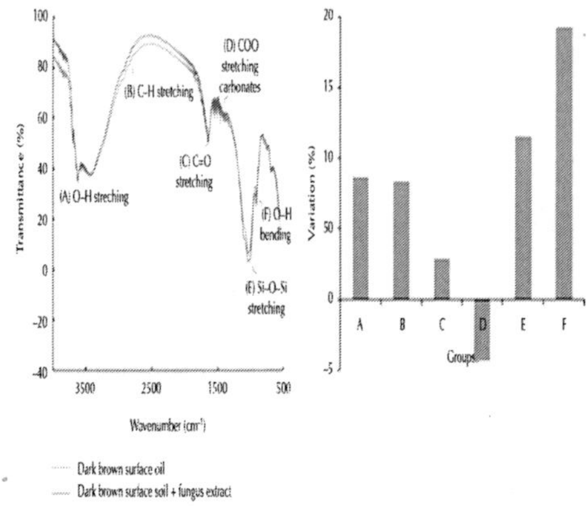

(a)

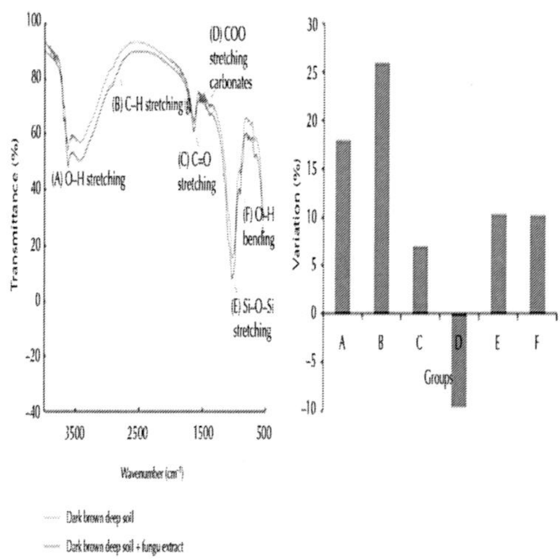

(b)

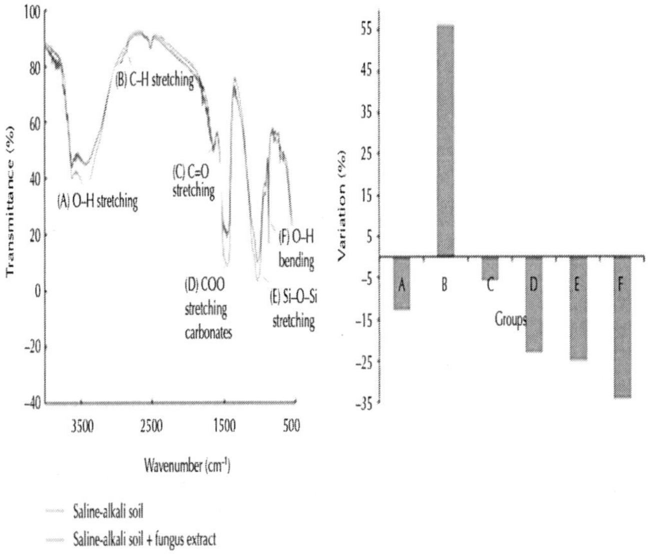

(c)

Figure 5: Infrared spectrum results after fungus extract addition. The labels are the same as those for Figure 4.

The addition of the fungus extract increased the relative crystallinity of soil colloids from both the surface and deep layers of dark brown forest soil (Figures 4(a) and 4(b)). Kaolinite increased by 300%, hydromica by 40–70%, and quartz by 83–157%. The grain size of hydromica and quartz both increased by 3–25%, but the kaolinite was reduced by 50–77%.

Unlike with dark brown forest soil, the addition of fungus extracts decreased the relative crystallinity of soil colloids from saline-alkali soil, except for quartz (a 17% increase) (Figure 4(c)). The largest reduction was 59% in kaolinite, while the decreases for hydromica, vermiculite, and calcite ranged from 15% to 25%. In most cases, the grain size of these soil minerals increased after the addition of the fungus extract, except for quartz (reduced 22%) (Figure 4(c)). Grain size increased by about 7% in vermiculite and calcite, while much larger increases (over 57%) were found in hydromica and kaolinite (Figure 4(c)).

IR Results

The addition of the fungus extract to soil colloids from the dark brown forest soil slightly reduced the amount of stretching of the COO^- and carbonate functional groups (about 10%) but increased O–H bending, increased stretching in most of the studied functional groups, including C–H, Si–O–Si, and O–H (all about 10–20%), and slightly increased (5%) C=O stretching (Figure 5(a)). Similar tendencies but slight differences in the size of changes were found in deep soil than in surface soil (Figure 5(b)).

Most functional group traits in the surface and deep layers of dark brown forest soil increased. However, a completely different pattern was found in saline-alkali soil (Figure 5(c)). In contrast with dark brown forest soil, the addition of fungus extracts to soil colloids from saline-alkali soil reduced the traits of most functional groups from 10% to 35% (Figure 5(c)). Functional group traits that decreased included O–H bending, C=O stretching, Si–O–Si stretching, O–H stretching, COO^- stretching, and carbonate stretching, with the exception of C–H stretching (a 56% increase) (Figure 5(c)).

XPS Results

Semiquantitative analysis of variable elements with and without the addition of the fungus extract was performed using XPS (Figure 6). In the case of soil colloids from the surface layer of dark brown forest soil, the addition of the fungus extract induced <5% changes in all elements, for example, a 5% increase in C1s and <5% decreases for all O1s, Si2p, N1s, and Ca2p (Figure 6(a)). Changes in variable elements in the deep soil due to the addition of fungus extracts were more evident than those in the surface layers (Figures 6(a) and 6(b)). The changes in C1s, O1s, and Si2p were less than 5%, while 6–9% decreases in N1s and Ca2p were observed (Figure 6(b)).

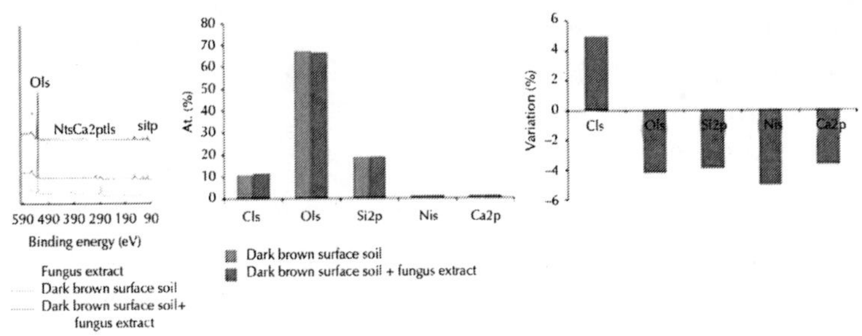

(a)

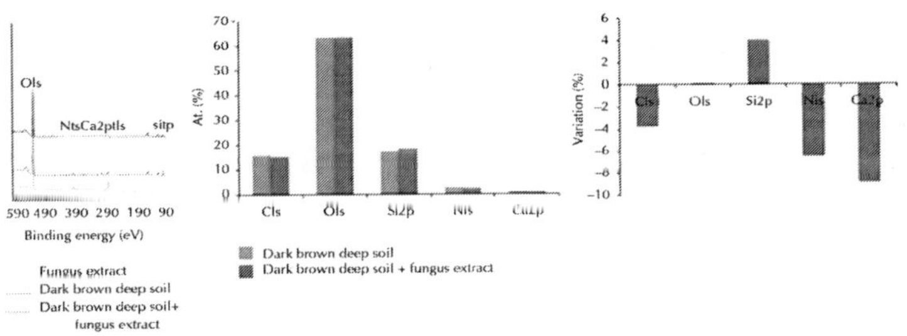

(b)

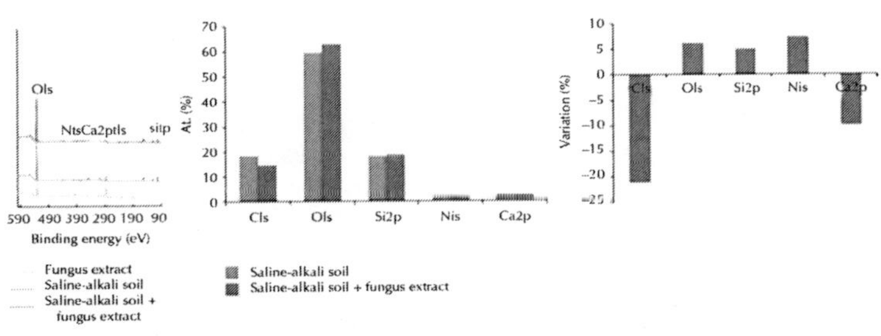

(c)

Figure 6: X-ray photoelectron spectroscopy results with and without fungus extract addition. The labels are the same as those for Figure 4.

Compared to the dark brown forest soil, addition of the fungus extract to the saline-alkali soil caused large reductions in variable elements (Figure 6(c)). C1s decreased by 21%, Ca2p by 10%, and O1s, Si2p, and N1s by 5%.

Stoichiometric changes induced by fungus extract addition were also found in the ratios among different elements (Table 1). In the case of the surface layer of dark brown forest soil, the ratios of C : N, Si : Ca, and C : Ca increased by 7–14%. In the case of the deep layer of dark brown forest soil, changes were also mainly found in C : N (6.5%), Si : Ca (20.5%), and C : Ca (17.1%). Stoichiometric changes were much more evident in saline-alkali soil than in dark brown forest soil. Over 25% decreases were found in C: O, C : N, and C : Si, and a 12.7% decrease was found in C : Ca. The Si: Ca ratio increased by 16.39% (Table 1).

Table 1: Results from X-ray photoelectron spectroscopy of the variation in element ratios with and without the addition of the fungus extract

	Dark brown forest soil—surface			Dark brown forest soil—deep			Saline-alkali soil		
	Control	Addition of the fungus extract	Variation%	Control	Addition of the fungus extract	Variation%	Control	Addition of the fungus extract	Variation%
C:O	0.16	0.18	9.58	0.25	0.24	-3.95	0.31	0.23	-25.68
C:N	7.30	8.10	10.96	6.30	6.50	3.17	12.00	8.80	-26.67
C:Si	0.57	0.62	8.77	0.90	0.84	-6.67	1.03	0.77	-25.24
Si:Ca	13.60	13.60	0.00	17.90	20.50	14.53	6.10	7.10	16.39
C:Ca	7.70	8.40	9.09	16.20	17.10	5.56	6.30	5.50	-12.70

DISCUSSION

Heavy soil degradation is common in China, and rehabilitation via vegetation recovery is mainly conducted in degraded regions, such as the saline-alkali soil region in the Songnen Plain [26]. Symbiotic associations between tree roots and ECM fungi play important roles in promoting the growth of these plants, and this promotion can be affected by variation in the strains of ECM fungi [46] and their interactions between biotic and abiotic factors [47]. ECM can aid the recovery of degraded soil both through the direct absorption of variable pollutants [17, 19] and indirect protection from rehabilitated vegetation [41]. In a previous study of ECM fungi, the main focus was on the relationship between ECM and plants. In the case of the ECM studied in this paper (Gomphidius viscidus), reviewed data have found significant increase in biomass of variable trees together with more soil nutrient absorption (Figure 7): that is, growth of biomass, ground diameter, lateral root and height of larch, pine, and oak increased from 20% to 40%, and much higher increase (50%–60%) in N, P absorption was found. This result indicates that soil nutrient absorption increase should be a basis for the biomass increase after ECM infection. However, few studies paid attention to the underlying mechanism of the interaction between ECM and the soil matrix. The findings in this paper showed that the ECM influences on soil colloids should be important aspect of degraded soil improvement.

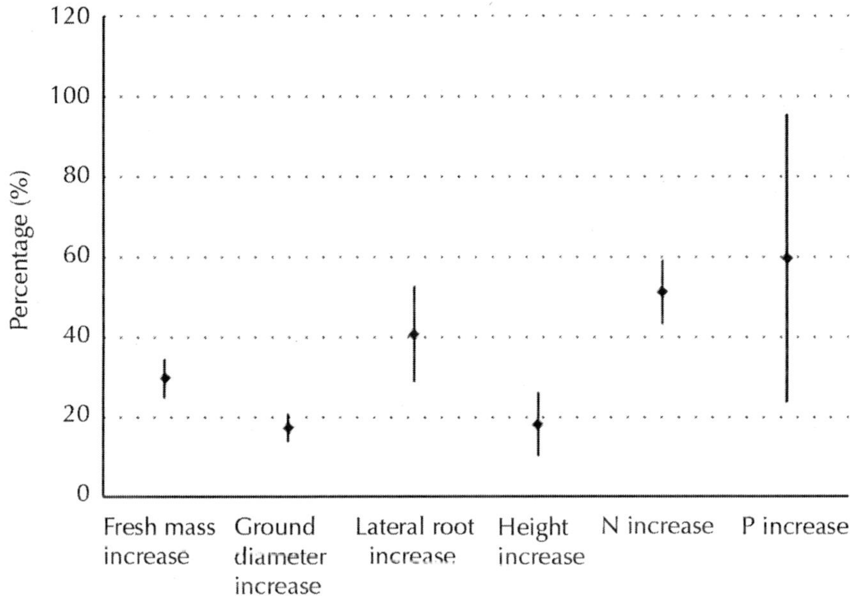

Figure 7: ECM infection influences on plant growth and biomass N, P absorption [39]. The ECM is the same species, Gomphidius viscidus, used in this study, and plant species are Larix kaempferi, Pinus tabulaeformis, and Quercus liaotungensis. The percentage in -axis is the improvement of each parameter relative to the untreated control (without ECM infection).

Dark brown forest soil and saline-alkali soil are two types of soil that are widespread in northeastern China. The first is an example of loam with good physicochemical properties [48], while the latter is notorious for poor physicochemical properties because of long-term degradation from human disturbance [26]. In the case of dark brown forest soil from surface and deep layers, both AFM and SEM images revealed that viscous materials wrapped around soil particles, filled some gaps among particles, and induced smoother surfaces with unclear edges (Figures 2(a)–2(d) and Figures 3(a)–3(d)). High absorption capacity is the basis for adhesive material absorption on soil colloids. This absorption has been reported previously [23]. Yan et al. reported that the maximum adsorption capacity (q_0) for fine soil colloids ranged from 169.6 to 203.7 μg mg^{-1} [49], which was higher than that for coarse soil colloids (81.0–94.6 μg mg^{-1}). Thus, physical absorption instead of chemical reactions possibly occurred in dark brown forest soil.

However, saline-alkali soil was different from dark brown forest soil. There were no clear adhesive layers on soil colloids, and smoother edges were found after the addition of fungus extracts (Figures 3(e) and 3(f)). Significantly larger particle sizes (Figure 1) and relatively loose interactions between different soil particles were observed in saline-alkali soil colloids (Figures 2(e) and 2(f)). Saline-alkali soil has a clay texture with tiny soil particles, high pH, and rich carbonates (especially calcium carbonate) [26]. This soil has poor soil physical structure with limited soil aggregates, and soil colloid surfaces are overloaded by Na^+ [48]. Wang et al. (2011) reported that the addition of the soil conditioner HPMA induced flocculation of soil colloids with looser structure and larger aggregates [26]. The findings on AFM, SEM images, and soil particle size indicate that the function of the fungus extract is like that of the soil conditioner (e.g., HPMA [26]), and a chemical reaction instead of physical adhesion may occur between the fungus extract and soil colloids. The mechanism of saline-alkali soil improvement by HPMA conditioner is to activate inorganic calcium in the soil via a chemical reaction between calcium carbonate and HPMA, as well as the exchange of Ca^{2+} and excessive Na^+on the surface of soil colloids [26]. A similar chemical reaction is possibly activated after the addition of the fungus extract to soil colloids from saline-alkali soils, owing to the fact that organic acids were present in the fungus extract solution.

It is worth mentioning that carbonate functional groups appear obviously reduced by 25%, perhaps because saline-alkali soil is basically an alkaline soil with a pH of 9-10 and includes more carbonate ($CaCO_3$) in calcite, while the fungus extract is generally acidic; therefore, combining them may cause some chemical reaction and lead to calcium carbonate being dispersed. Dissolution of calcium carbonate could lead to more Ca^{2+} being adsorbed on the surface of soil colloids, which would tend to increase soil colloid dispersion and form a more loose soil colloid apparent structure [26], as shown in Figure 2.

Measuring differences in mineral composition on the surface of soil colloids and in mineral crystallinity may help explain the changes in particle diameter and surface structure as shown in Figures 1–3. The techniques of XRD, IR, and XPS are commonly used for mineral crystallinity observation [42, 43], functional group identification in soils [44], and elemental composition changes [50]. These methods could be used to examine mineral crystallinity changes, variable functional group changes, and the elemental composition of soil colloids, together with surface and particle size changes after the addition of the fungus extract.

XRD data proved that fungus extract treatment could change the relative crystallinity and grain size of variable soil minerals and showed differences between dark brown forest soil and saline-alkali soil (Figure 4). In the case of dark brown forest soil, addition of fungus extracts 30%~400% increased crystallinity of kaolinite, hydromica, and quartz, while 17%~58% decreases were found in saline-alkali soil (except for a 15% increase for quartz) (Figure 4). These data were well-matched with the observations from SEM and AFM; that is, soil colloid particle edges became blurred with covering materials in dark brown forest soil, while soil colloids from saline-alkali soil tended to be dispersed with relatively loose interactions (Figures 2–3). By using a similar XRD method, Zheng and Zhao (2011) found that soil clay minerals (illite, kaolinite, and montmorillonite) are very sensitive to disturbance, and their changes could affect soil fertility [51]. Our results showed that ECM fungi in soil may favor this process via their effect on the relative crystallinity and grain size of soil minerals.

IR can provide detailed information about variable functional groups [44], and IR data from this study showed ECM fungus could induce evidently different responses between saline-alkali soil and dark brown forest soil. Traits of functional groups such as O–H stretching and bending, C–H stretching, and Si–O–Si stretching in soil colloids of dark brown forest soil were increased by 8–27% after the addition of the fungus extract (Figures 5(a) and 5(b)). However, all the functional group traits in saline-alkali soil were reduced by about 10–35%, except for C–H stretching. The changes in functional group composition (Figure 4) and mineral crystallization (Figure 5) in soil colloids were accompanied by relative changes in element composition (Figure 6 and Table 1), too. Compared with very slight changes in dark brown forest soil, addition of the fungus extract to degraded soil (saline-alkali soil) majorly reduced C1s (21%) and made the C : O, C : N, C : Si, and C : Ca element ratios more similar to those of high-quality dark brown forest soil (Table 1). ECM inoculum has been commonly used during forestry plantings and native prairie restorations to enhance tree and plant growth [11].

The present study clarifies the mycorrhizal-soil interaction. Field studies indicate that mycorrhizal inocula benefit native plant production and establishment in severely degraded areas [11, 52], and commercial mycorrhizal-soil conditioners are now available for vegetation recovery [7]. The interactions between ECM fungus extracts and soil colloids observed in this study indicate that there are relatively fast physical and chemical reactions between fungi and tiny soil particles and that the clarification of this process may promote the understanding of the function of ECM in soil nutrient exploitation and soil health maintenance.

CONCLUSIONS

The addition of fungus extract to soil colloids could significantly affect the surface structure, mineral crystallization, functional groups, and elemental composition of colloids, although the effects were different for dark brown forest soil (a good loam) and saline-alkali soil (a heavily degraded soil). For dark brown forest soil, physical absorption at the surface of soil colloids together with increases in functional group traits such as O–H stretching and bending, C–H stretching, C=O stretching, and Si–O–Si stretching and the relative crystallinity of kaolinite, hydromica, and quartz were observed after the addition of fungus extracts. In the case of degraded saline-alkali soil, the particle diameter and grain size of variable soil minerals were sharply increased, and remarkable reductions were found in the relative crystallization of variable minerals and most functional groups (5 out of 6), showing that chemical reactions instead of physical absorption possibly occurred after the addition of the fungus extract. As good loams and degraded soils are typical in northeastern China, the findings in this paper will improve the understanding of the mechanisms for ECM-aided soil improvements, especially for highly degraded soil.

ACKNOWLEDGMENTS

This study was supported financially by China's Ministry of Science and Technology (2011CB403205), China's National Foundation of Natural Sciences (31170575), China's Postdoctoral Foundation (201003406 and 20080430126), and Basic Research Fund for National Universities from Ministry of Education of China (DL12DA03).

REFERENCES

1. R. Lal, "Soil degradation by erosion," Land Degradation & Development, vol. 12, no. 6, pp. 519–539, 2001.
2. S. Kapur and E. Akca, "Global assessment of land degradation," in Encyclopedia of Soil Science, R. Lal, Ed., pp. 296–306, Marcel Dekker, New York, NY, USA, 2002.
3. C. Baum, K. Hrynkiewicz, P. Leinweber, and R. Meißner, "Heavy-metal mobilization and uptake by mycorrhizal and nonmycorrhizal willows (Salix × dasyclados)," Journal of Plant Nutrition and Soil Science, vol. 169, no. 4, pp. 516–522, 2006.
4. K. Hrynkiewicz, C. Baum, and P. Leinweber, "Mycorrhizal community structure, microbial biomass P and phosphatase activities under Salix polaris as influenced by nutrient availability," European Journal of Soil Biology, vol. 45, no. 2, pp. 168–175, 2009.
5. A. G. Khan, C. Kuek, T. M. Chaudhry, C. S. Khoo, and W. J. Hayes, "Role of plants, mycorrhizae and phytochelators in heavy metal contaminated land remediation," Chemosphere, vol. 41, no. 1-2, pp. 197–207, 2000.
6. X. M. Hua, Trees Mycorrhizal Research, Chinese Science and Technology Press, Beijing, China, 1995.
7. K. M. Vogelsang, J. D. Bever, M. Griswold, and P. A. Schultz, "The use of mycorrhizal fungi in erosion control applications," Final Report for Caltrans 65A0070, California Department of Transportation, Sacramento, Calif, USA, 2004.
8. Y. Li, S. Q. Wang, and W. P. Ju, "Effects of mycorrhizal fungi on pine seedling root disease prevention and cure," Journal of Liaoning Forestry Science and Technology, vol. 4, pp. 21–22, 2000.
9. X. F. Yan and Q. Wang, "Effects of ectomycorrhizal inoculation on the seeding growth of QuercusLiaotungensis," Acta Phytoecologica Sinica, vol. 26, no. 6, pp. 701–707, 2002.

10. Q. Ma, J. G. Huang, and J. B. Jiang, "Effect of inoculating with the ectotrophic mycorrhizal epiphyte on the Pinus massoniana seedling growth," Journal of Fujian Forestry Science and Technology, vol. 32, no. 2, pp. 85–88, 2005.
11. B. M. Ohsowski, J. N. Klironomos, K. E. Dunfield, and M. Harta Miranda, "The potential of soil amendments for restoring severely disturbed grasslands," Applied Soil Ecology, vol. 60, pp. 77–83, 2012.
12. D. J. Read, "Mycorrhizas in ecosystems," Experientia, vol. 47, no. 4, pp. 376–391, 1991.
13. N. L. Carly, E. D. Laura, and H. Jonathan, "Mycorrhizae and soil phosphorus affect growth of Celastrus orbiculatus," Biological Invasions, vol. 13, no. 10, pp. 2339–2350, 2011.
14. K. Bojarczuk and B. Kieliszewska-Rokicka, "Effect of ectomycorrhiza on Cu and Pb accumulation in leaves and roots of silver birch (Betula pendula Roth.) seedlings grown in metal-contaminated soil,"Water, Air and Soil Pollution, vol. 207, no. 1–4, pp. 227–240, 2010.
15. C. Voiblet, S. Duplessis, N. Encelot, and F. Martin, "Identification of symbiosis-regulated genes inEucalyptus globulus—Pisolithus tinctorius ectomycorrhiza by differential hybridization of arrayed cDNAs," The Plant Journal, vol. 25, no. 2, pp. 181–191, 2001.
16. A. Au ina, M. Rudawska, T. Leski, D. Ryliškis, M. Pietras, and E. Riepšas, "Ectomycorrhizal fungal communities on seedlings and conspecific trees of Pinus mugo grown on the coastal dunes of the Curonian Spit in Lithuania," Mycorrhiza, vol. 21, no. 3, pp. 237–245, 2011.
17. K. Hrynkiewicz, G. Dabrowska, C. Baum, K. Niedojadlo, and P. Leinweber, "Interactive and single effects of ectomycorrhiza formation and Bacillus cereus on metallothionein MT1 expression and phytoextraction of Cd and Zn by willows," Water, Air & Soil Pollution, vol. 223, no. 3, pp. 957–968, 2012.
18. T. Leski and M. Rudawska, "Ectomycorrhizal fungal community of naturally regenerated European larch (Larix decidua) seedings," Symbosis, vol. 56, no. 2, pp. 45–53, 2012.
19. D. Zimmer, C. Baum, P. Leinweber, K. Hrynkiewicz, and R. Meissner, "Associated bacteria increase the phytoextraction of

cadmium and zinc from a metal-contaminated soil by mycorrhizal willows,"International Journal of Phytoremediation, vol. 11, no. 2, pp. 200–213, 2009

20. G. K. Podila, J. Zheng, S. Balasubramanian et al., "Fungal gene expression in early symbiotic interactions between Laccaria bicolor and red pine," Plant and Soil, vol. 244, no. 1-2, pp. 117–128, 2002.
21. A. L. Quéré, D. P. Wright, B. Söderström, A. Tunlid, and T. Johansson, "Global patterns of gene regulation associated with the development of ectomycorrhiza between birch (Betula pendula Roth.) and Paxilus involutus (Batsch) fr," Molecular Plant-Microbe Interacionst, vol. 18, no. 7, pp. 659–673, 2005.
22. G. Heller, A. Adomas, G. Li et al., "Transcriptional analysis of Pinus sylvestris roots challenged with the ectomycorrhizal fungus Laccaria bicolor," BMC Plant Biology, vol. 8, article 19, 2008.
23. D. M. Zhou, D. J. Wang, L. Cang, X. Hao, and L. Chu, "Transport and re-entrainment of soil colloids in saturated packed column: effects of pH and ionic strength," Soils Sediments, vol. 11, no. 3, pp. 491–503, 2011.
24. N. M. DeNovio, J. E. Saiers, and J. N. Ryan, "Colloid movement in unsaturated porous media: recent advances and future directions," Vadose Zone Journal, vol. 3, no. 2, pp. 338–351, 2004.
25. C. D. Barton and A. D. Karathanasis, "Influence of soil colloids on the migration of atrazine and zinc through large soil monoliths," Water, Air and Soil Pollution, vol. 143, no. 1–4, pp. 3–21, 2003.
26. W. J. Wang, H. S. He, Y. G. Zu et al., "Addition of HPMA affects seed germination, plant growth and properties of heavy saline-alkali soil in Northeastern China: comparison with other agents and determination of the mechanism," Plant and Soil, vol. 339, no. 1, pp. 177–191, 2011.
27. C. C. Muggler, T. H. Pape, and P. Buurman, "Laser grain-size determination in soil genetic studies. 2. Clay content, clay formation, and aggregation in some Brazilian oxisols," Soil Science, vol. 162, no. 3, pp. 219–228, 1997.
28. L. Beuselinck, G. Govers, and J. Poesen, "Assessment of micro-aggregation using laser diffractometry,"Earth Surface Processes and Landforms, vol. 24, no. 1, pp. 41–49, 1999.

29. T. M. Zobeck, "Rapid soil particle size analyses using laser diffraction," Applied Engineering in Agriculture, vol. 20, no. 5, pp. 633–639, 2004.
30. J. R. Campbell, "Limitations in the laser particle sizing of soils," in Advances in Regolith, pp. 38–42, 2003.
31. S. Fontaine, S. Barot, P. Barré, N. Bdioui, B. Mary, and C. Rumpel, "Stability of organic carbon in deep soil layers controlled by fresh carbon supply," Nature, vol. 450, no. 8, pp. 277–280, 2007.
32. B. D. Lee, S. K. Sears, R. C. Graham, C. Amrhein, and H. Vali, "Secondary mineral genesis from chlorite and serpentine in an ultramafic soil toposeqnence," Soil Science Society of America Journal, vol. 67, no. 4, pp. 1309–1317, 2003.
33. G. Haberhauer and M. H. Gerzabek, "FTIR-spectroscopy of soils-characterisation of soil dynamic processes," Trends in Applied Spectroscopy, vol. 3, pp. 103–109, 2001.
34. C. T. Johnston and Y. O. Aochi, "Fourier transform infrared and raman spectroscopy," in Methods of Soil Analysis. Part 3, J. M. Bartels and J. M. Bigham, Eds., pp. 269–321, Soil Science Society of America, American Society of Agronomy, Madison, Wis, USA, 1996.
35. R. J. Cox, H. L. Peterson, J. Young, C. Cusik, and E. O. Espinoza, "The forensic analysis of soil organic by FTIR," Forensic Science International, vol. 108, no. 2, pp. 107–116, 2000.
36. M. Soma, H. Seyama, N. Yoshinaga, B. K. G. Theng, and C. W. Childs, "Bonding state of silicon in natural ferrihydrites by X-ray photoelectron spectroscopy," Clay Science Society of Japan, vol. 9, no. 6, pp. 385–391, 1996.
37. C. W. Childs, K. Inoue, H. Seyama, M. Soma, B. K. G. Theng, and G. Yuan, "X-ray photoelectron spectroscopic characterization of silica springs allophane," Clay Minerals, vol. 32, no. 4, pp. 565–572, 1997
38. G. Yuan, M. Soma, H. Seyama, B. K. G. Theng, L. M. Lavkulich, and T. Takamatsu, "Assessing the surface composition of soil particles from some Podzolic soils by X-ray photoelectron spectroscopy,"Geoderma, vol. 86, no. 3-4, pp. 169–181, 1998.
39. Y. H. Li, B. Y. Zhang, Y. Jiang, and W. J. Wang, "The effect of organic and inorganic carbon on extracellular enzyme activity

of acid phosphatase and proteases in three kinds of fungal hyphae,"Bulletin of Botanical Research, vol. 33, no. 4, pp. 404–409, 2013.

40. Y. H. Dong, C. Hao, Y. R. Niu, and Q. J. Chen, "The ecological elementary research ofchroogomphisrutillus," Chinese Agricultural Science Bulletin, vol. 26, no. 7, pp. 191–194, 2010.

41. F. Q. Yu and P. G. Liu, "Reviews and prospects of the ectomycorrhizal research and application," Acta Ecologica Sinica, vol. 22, no. 22, pp. 2217–2226, 2002.

42. B. Xu, J. M. Man, and C. X. Wei, "Methods for determining relative crystallinity of plant starch X-ray powder diffraction spectra," Chinese Bulletin of Botany, vol. 47, no. 3, pp. 278–285, 2012.

43. P. R. Xie, Soil Chemistry Mineralogy Properties in Northeast of China, Science Press, Beijing, China, 2010.

44. P. Marc and G. E. Jacques, Handbook of Soil Analysis, Springer, Berlin, Germany, 2006.

45. D. L. Spark, Methods of Soil Analysis Part 3: Chemical Methods, Soil Science Society of America, 1996.

46. S. E. Smith and D. J. Read, Mycorrhizal Symbiosis, Academic Press, London, UK, 1997.

47. C. Baum, Y. K. Toljander, K. U. Eckhardt, and M. Weih, "The significance of host-fungus combinations in ectomycorrhizal symbioses for the chemical quality of willow foliage," Plant and Soil, vol. 323, no. 1-2, pp. 213–224, 2009.

48. HLJTR. Soil management bureau and soil census office of Heilongjiang Province, China. Soil of Heilongjiang Province. China Agriculture, Beijing, China, 1993.

49. J. L. Yan, G. X. Pan, L. Q. Li, G. Quan, C. Ding, and A. Luo, "Adsorption, immobilization, and activity ofβ-glucosidase on different soil colloids," Journal of Colloid and Interface Science, vol. 348, no. 2, pp. 565–570, 2010.

50. B. Stenberg, R. A. ViscarraRossel, A. M. Mouazen, and J. Wetterlind, Visible and Near Infrared Spectroscopy in Soil Science, Academic Press, Burlington, Mass, USA, 2010.

51. Q. F. Zheng and L. P. Zhao, "Effect of agricultural use on clay minerals and nutrient of black soil in Northeast of China," Journal of Jilin Agricultural Sciences, vol. 36, no. 5, pp. 29–32, 2011.

52. S. R. Matias, M. C. Pagano, F. C. Muzzi et al., "Effect of rhizobia, mycorrhizal fungi and phosphate-solubilizing microorganisms in the rhizosphere of native plants used to recover an iron ore area in Brazil," European Journal of Soil Biology, vol. 45, no. 3, pp. 259–266, 2009.

Chapter 5

Coupled Electrokinetics-Adsorption Technique for Simultaneous Removal of Heavy Metals and Organics from Saline-Sodic Soil

Salihu Lukman[1] Mohammed Hussain Essa[1] Nuhu Dalhat Mu'azu[2], and Alaadin Bukhari[1]

[1]Department of Civil and Environmental Engineering, King Fahd University of Petroleum and Minerals, Dhahran 31261, Saudi Arabia

[2]Environmental Engineering Department, University of Dammam, Dammam 31451, Saudi Arabia

ABSTRACT

In situ remediation technologies for contaminated soils are faced with significant technical challenges when the contaminated soil has low permeability. Popular traditional technologies are rendered ineffective due to the difficulty encountered in accessing the contaminants as well as when employed in settings where the soil contains mixed contaminants such as petroleum hydrocarbons, heavy metals, and polar organics. In this study, an integrated in situ remediation technique that couples electrokinetics with adsorption, using locally produced granular activated carbon from date palm pits in the treatment zones that are installed directly to bracket the contaminated soils at bench-scale, is investigated. Natural saline-sodic soil, spiked with contaminant mixture (kerosene, phenol, Cr, Cd, Cu, Zn, Pb, and Hg), was used in this study to investigate the efficiency of contaminant removal. For the 21-day period of continuous electrokinetics-adsorption experimental run, efficiency for the removal of Zn, Pb, Cu, Cd, Cr, Hg, phenol, and kerosene was found to reach 26.8, 55.8, 41.0, 34.4, 75.9, 92.49, 100.0, and 49.8%, respectively. The results obtained suggest that integrating adsorption into electrokinetic technology is a promising solution for removal of contaminant mixture from saline-sodic soils.

INTRODUCTION

Rapid proliferating industrialization has been recorded in recent decades. One of the major environmental consequences of these progressive achievements is the improper release of elevated amounts of variety of organic and inorganic pollutants into the environment. These pollutants could enter the environment directly as a result of accidents, spills during transportation, and leakage from waste disposal sites, storage sites, industrial facilities, and so forth, thereby contaminating the environment. Cooccurrence of complex chemical mixtures such as total petroleum hydrocarbons (TPH), phenols, heavy metals (such as Cr, Cd, Cu, Zn, Pb, and Hg), radionuclides, and pesticides at remediation sites pose potential dangers to human health and the environment and further complicate the remediation process. Some of the pollutants encountered in contaminated soils may be treated using processes like biodegradation, vapor extraction, chemical oxidation, thermal desorption, and incineration. Satisfactory results may not be obtained when these treatment processes are applied to low-permeability soils or those sites contaminated by mixed contaminants (inorganic and organic wastes) because of the difficulty in accessing the contaminants to uniformly deliver treatment reagents. Therefore, innovative remediation technologies must be developed to study the in situ removal of contaminant mixture from soil to ensure a sustainable environment. This has given birth to an important area of current research in in situ soil remediation technology [1–4]. In situ treatment technologies for contaminated soils and groundwater have been the subject of a great deal of research in the last three decades owing to their attendant advantages: potential lower cost, less environmental disruption, and reduction in worker exposure to hazardous materials [1].

Saline-sodic soils (usually found in arid and semiarid regions) possess high electrical conductivity (>4 dS/m) which prevents the application of appropriate voltage gradient in an electrokinetic study owing to current limitations. In addition, these soils are associated with high pH > 8.2, dominated by 2 : 1 type clay minerals and exchangeable sodium at high levels greater than 15 [5, 6]. These properties make saline-sodic soils to be high acid buffering, alkaline, and very difficult to remediate if they are contaminated with heavy metals due to precipitation concerns in alkaline environment. These extreme soil characteristics pose great difficulty in having such soils remediated from mixed contaminants using electrokinetic-based technique. Despite these challenges posed by the soil characteristics, there is a need to investigate possible remediation of such soils using the integrated electrokinetics-adsorption technique giving its promise in remediating low permeability soils [7]. The usual voltage gradient of 1 V/cm for bench-scale studies when applied to such soils could lead to high electric current flow. This in turn could lead to excessive soil heating, reduction in the soil moisture content, high energy and process fluid consumption, and in some cases higher percentage removal of contaminants.

It has been observed that contaminated soils do not contain single contaminants. Several pollutants appear in the soil as mixed components. In reality, soil polluted with organic contaminants often contains other contaminants such as heavy metals. The implication of the presence of the different nature of the two contaminant groups is that there may be synergistic or antagonistic effects on their respective removal using electrokinetic remediation technique [8, 9]. Reddy [10] posited that the presence of mixed contaminants will retard individual contaminant migration and removal. Also, as organic pollutants are removed by electroosmotic flow and heavy metals by electromigration, the solubility as well as hydrophobicity disparities between the organic pollutants and heavy metals indicates the complexity of electrokinetic remediation of soils polluted with mixed contaminants. To date, several studies have been conducted using electrokinetics (EK) for mixed contamination [11–14]. Of the several electrokinetic remediation techniques, Lasagna process has been found to yield the best removal efficiency of organic contaminants from soils [15]. The general concept of the Lasagna process is the transportation of contaminants from contaminated soil section into treatment zones using major electrokinetic transport mechanisms (i.e., electroosmosis or electromigration). Once at the treatment zones, the contaminants may be removed from the pore water by sorption, degradation, or immobilization depending on treatment zone design [1, 16–18]. Detailed studies of all previous works on the Lasagna process which span from bench-scale investigations to full field-scale remediation of contaminated soils have been reported elsewhere [1, 7, 16–23]. Lasagna process usually uses activated carbon as the sorbent material to improve the removal of contaminants from contaminated soil [1, 22, 23].

The main aim of this study is to investigate the possible application of the coupled electrokinetics-adsorption innovative technique that combines electrokinetics and adsorption using locally produced granular activated carbon (GAC) from date palm pits for remediation of local saline-sodic soil contaminated with mixture of toxic pollutants comprising of petroleum byproduct (kerosene), organic compound (phenol), and heavy metals (Cr, Cu, Cd, Zn, Pb, and Hg).

MATERIALS AND METHODS

Characterization

Clay used in this study is a local Saudi Arabian clay from Al-Hassa oasis. The clay pH, moisture content, soil organic matter (SOM), electrical conductivity, surface area, and elemental analysis using scanning electron microscopy (SEM) and X-ray diffraction (XRD) methods were determined according to the protocol outlined in the American Society of Testing and Materials (ASTM) standards [25] and reported elsewhere [24]. These physicochemical properties of the saline-soil are presented in Table 1. The mineralogical composition obtained from the XRD result shows that the soil consists mainly of montmorillonite, quartz, and calcite minerals. Scanning electron microscopy revealed the dominant elements to be O and Si whose percentage compositions are 53.11% and 18.46%, respectively. Other elements present in lesser quantities are Ca (7.09%), Al (6.76%), Fe (5.20%), Na (2.69%), Cl (2.69%), K (2.29%), and Mg (1.73%). The GAC used in the present study was produced locally from date palm pits as described elsewhere [26, 27].

Table 1: Physicochemical properties of saline-sodic soil [24]

Property	Value
pH (ASTM D 4972)	9.00
Moisture content, % (ASTM D 2216)	3.91
Soil organic matter, % (ASTM D 2974)	3.26
Hydraulic conductivity, cm/s (ASTM D 5084)	6.91× 10-09
Electrical conductivity, dS/m (ASTM D 1125)	8.62
Average pore width (by BET*), Å	75.64
BET specific surface area, m2/g (UOP964)	42.13
Pore volume, cm3/g (UOP964)	0.08
Specific gravity (ASTM D 854)	2.77
Liquid limit (ASTM D 4318)	44.71
Plastic limit (ASTM D 4318)	25.95
Plasticity index	18.76
USCS classification (ASTM D 2487)	CL (Clay)
Particle size distribution (ASTM D 422)	
Clay, %	78
Silt, %	6
Very fine sand, %	10
Fine sand, %	5
Medium sand, %	1

*BET: Brunauer-Emmett-Teller.

Adsorption Testing

Single and competitive adsorption of five heavy metals (Cr, Cd, Cu, Zn, and Pb) was performed to determine the selectivity sequence and to understand the adsorption behavior of these metals under different pH conditions. This is particularly important to this study, because, soil mineralogy affects heavy metal adsorption behavior and selectivity sequence under different pH conditions. Lukman et al. [24] reported the detailed procedures carried out for the competitive adsorption testing.

Coupled Electrokinetics-Adsorption Study

A total of three bench-scale experiments were performed to investigate the treatability of the contaminant mixture using the coupled electrokinetics-adsorption technique and to understand the operating peculiarities of the saline-sodic soil. Two of the experiments had GAC chambers bracketing the contaminated soil chamber and were operated at voltage gradients of 0.6 V/cm and 1 V/cm. The third one utilized only electrokinetics at voltage gradient of 0.6 V/cm.

Reactor Design and Experimental Procedures

The Plexiglass reactor total volume was about 2268 cm^3, made of seven chambers. The overall reactor dimensions are: 24 cm (long) × 10 cm (width) × 12 cm (depth). Approximately 1 kg of local KSA soil was artificially spiked with kerosene, heavy metals (Cu, Cr, Cd, Pb, Zn, and Hg), and phenol at predetermined concentrations. Thorough mixing was done using mechanical mixer (Gilson Company, Inc.) so as to achieve a homogeneous distribution of the contaminants around the soil matrix. The mixed spiked soil was placed in a fume hood for drying over a period of time necessary to evaporate the solvents (hexane and distilled water). Distilled water was added to adjust the final moisture content of the soil to about 33–70%. The initial conditions of the soil pH, moisture content, organic matter, and electrical conductivity were measured as well as the actual initial concentrations of the contaminants. Then, the uniformly mixed contaminated soil was placed into the cell layer by layer. Each layer was compacted with stainless steel spatula, so that the amount of void space was minimized. The Lasagna cell used for the experiments consists of the cell, two graphite electrodes serving as anode and cathode, DC power supply (LG, GP-505), processing fluid reservoirs, heavy duty recirculation pump (BVP Instratec), portable data logger (TDS-303, Tokyo Sokki Kenkyujo Co., ltd) for real-time monitoring of temperature, electric current, and voltage across the system (Figure 1). The two electrode compartments with 240 mL working volume, placed at each end of the cell, were isolated from the soil zone by a porous Perspex plate and filter paper. The conditioning of the electrolyte was controlled by circulation of the anolyte (2 N NaOH) and catholyte (1 N HNO_3) using pump that is directly attached to the electrode compartments. Two planar-shaped electrodes, 10 cm × 10 cm × 0.5 cm, were used to generate a uniform electric field. Within the described cell, two treatment zones that cut across the cell vertically bracketing the spiked soil compartment were filled with the GAC. The data monitoring system was recording electric current variation, applied voltage, and temperature of the soil compartments on-line following a 30-minute preset time step and automatically stores them for subsequent retrieval using floppy disc which can be read using personal computer for easy data and energy consumption analysis. Treatment without GAC that is using only electrokinetics was conducted so as to ascertain the contribution of the GAC in contaminant removal. The power supply provides a constant DC electric voltage for the electrokinetic test. Every week, fractions of the soil specimens were taken to determine the residual concentrations of the contaminants, soil pH, water content, organic matter, and electrical conductivity. Upon the completion of each test, the electrode assemblies were disconnected, and the soil specimen was extruded from the cell, sectioned into parts, weighed, and preserved in glass vials for organic extraction, heavy metal digestion, and subsequent analyses using the analytical procedures outlined below.

Figure 1: Coupled electrokinetics-adsorption experimental setup.

Analytical Procedures for Contaminant Extraction and Analysis

Heavy Metals. Extraction of heavy metals from soil samples was performed according to guidelines spelt out in EPA Method 3050B for acid digestion of soils, sediments, and sludges [28] and analyzed using flame atomic absorption spectrometry (AAnalyst 700, PerkinElmer). All soil samples were extracted in duplicates. EPA Method 7000B [29] was employed for heavy metal analysis using flame atomic absorption spectrometry except for mercury which was analyzed using mercury analyzer (Solid Mercury Analyzer SMS 100, Perkin Elmer) according to EPA Method 7473 [30].

Kerosene and Phenol. A mixture of methylene chloride and hexane (1 : 1) (v/v) was used as the extraction solvent. Soil samples were extracted using pressurized fluid extraction according to EPA Method 3545 procedures [31] using accelerated solvent extractor (ASE 200, Dionex). No separate extraction procedure is required for phenol, because, it will also be extracted along with kerosene from the soil sample. Volume of extract generated was then injected into the GC-MS (Clarus 580, PerkinElmer) equipped with autosampler for analysis. TPH quantification was done by using the total chromatographic area counts using retention time range for the elution of hydrocarbon within the kerosene range C_8-C_{16}Guidelines spelt out in EPA Method 8270D [32] for the quantification of semivolatile organics by GC-MS were adhered to.

Testing Program

Overall, three experiments were conducted using an initial intended concentration of 100 mg/kg for each contaminant. Three designations used are EK-GAC-1, EK-GAC-2, and EK for coupled EK, adsorption 1 and 2, and EK, respectively. EK-GAC-1 utilized 1 V/cm and served to produce some preliminary information such as the feasibility of attaining reasonable percentage removal and processing fluid conditioning requirements. EK-GAC-2 and EK were then run simultaneously at 0.6 V/cm to ascertain the preference of using the integrated approach over EK alone.

RESULTS AND DISCUSSION

Clay and GAC Characteristics

From Table 1, the soil pH (9.00) indicates that it contains appreciable soluble salts capable of undergoing alkaline hydrolysis such as sodium carbonate [6]. The hydrolysis of calcite ($CaCO_3$) may be limited by its low solubility, thus producing a pH of about 8–8.2 in soils. In addition, Na^+ ions do not strongly compete with H^+ ions for exchange sites as does Ca^+ ions which are strongly and more tightly held on the soil surface. The inability of the displaced Na^+ ions to inactivate OH^- ions results in increased soil pH, which is usually greater than 8.2. Similarly for soil whose pH is greater than 8.2, its exchangeable sodium percentage is greater than 15 [6]. Presence of calcite coupled with alkaline hydrolysis of sodium carbonate gives high electrical conductivity to the soil (8.62 dS/m). The role of soil organic matter (SOM) in heavy metal adsorption is not to be underrated by its low value (3.26%). This is because of the high specific surfaced area and cation exchange capacity possessed by SOM which may reach up to 800–900 m^2 /g and 150–300 cmol/kg, respectively [33]. The physicochemical properties and morphological characteristics of the clay and GAC are detailed elsewhere [24, 26].

Single and Competitive Adsorption of Heavy Metals on Clay

Lukman et al. [24] have found out that the adsorptive capacities of Cu and Zn ions are higher in the multicomponent adsorption scenario than in the single component scenario as presented in Figures 2 and 3. The adsorption selectivity sequences obtained using the coefficient of distribution for the single and multicomponent scenarios are Cr > Pb > Cu > Cd > Zn and Cr > Cu > Pb > Cd > Zn, respectively [24]. Srivastava et al. [34] have reported similar selectivity sequence for the multicomponent scenario. Yong et al. [35] have identified the general factors that influence selectivity sequence to be ionic size or activity, first hydrolysis constant, soil type, and pH of the system. From the multicomponent desorption study presented in Figure 4, it can inferred that trivalent Cr ions were tightly held by the soil surface, thus having the least percentage desorption, followed by Cd and Cu ions. Reddy and his coworkers [36–38] have reported that trivalent Cr ions adsorb highly to soil solids and form cationic species that are insoluble over a wide range of pH. This is in line with the present findings (Figures 3 and 4) which revealed high selectivity for the trivalent Cr during multicomponent adsorption and desorption tests. It may be argued that the solubility of heavy metal ions at alkaline pH is very low due to their precipitation as insoluble hydroxides. However, in this study, the main focus is not on the adsorbed or precipitated species but rather on the mobile or dissolved species that can be removed via electromigration or electroosmotic flow during electrokinetic remediation. Hence, percentage removal (Figures 2 and 3) refers to the amount adsorbed by the soil minerals plus any precipitated metal species.

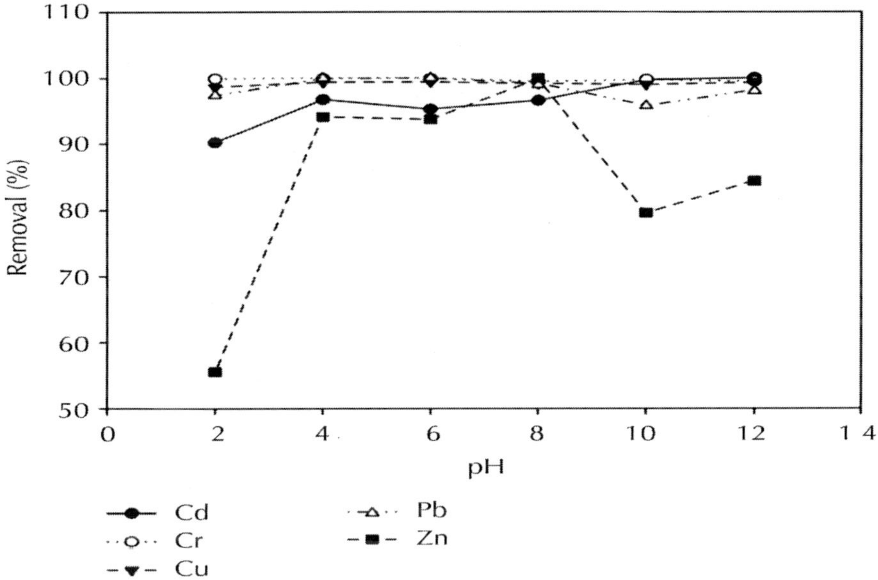

Figure 2: Percentage of heavy metals adsorbed plus any precipitated species at different pH for single component adsorption scenario [24].

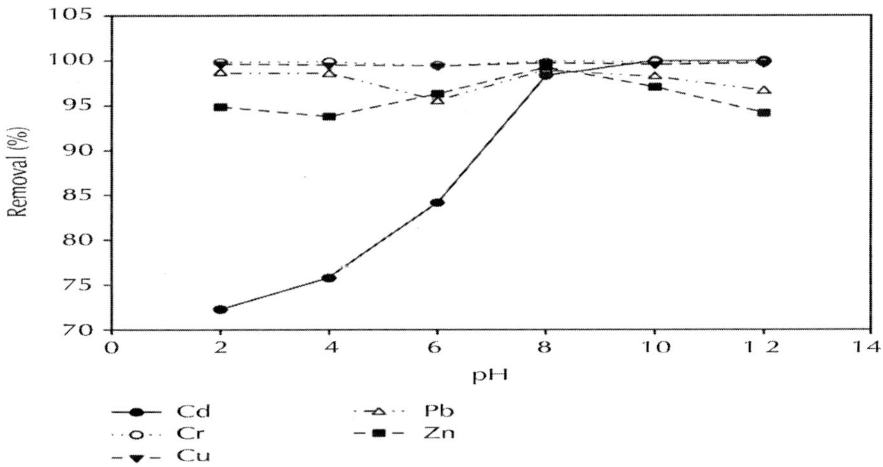

Figure 3: Percentage of heavy metals adsorbed plus any precipitated species at different pH for multicomponent adsorption scenario [24].

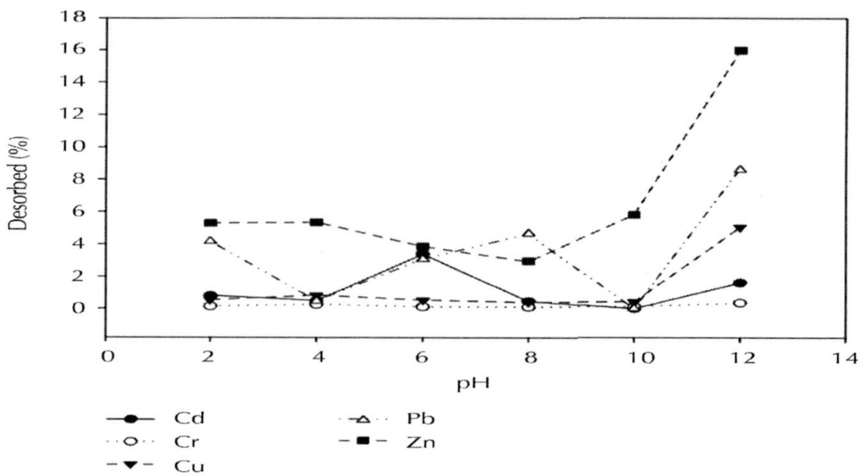

Figure 4: Percentage of heavy metals desorbed plus any precipitated species at different pH for multicomponent desorption scenario [24].

Soil pH Distribution during Coupled Electrokinetics-Adsorption Remediation

The pH value of the investigated clay is naturally alkaline (pH = 9), which promotes heavy metals precipitation and adsorption onto the clay surface depending on the metal speciation. At the end of the 21-day period, the pH distribution within the soil was found to be approximately 12 (Figure 5) for all the three runs despite dissimilar application of voltage gradients which might be expected to increase the rate of production of H^+ and OH^- radicals and their subsequent migration to the opposite electrodes for higher voltage gradient. This high pH environment might be explained by the presence of calcite in the soil minerals which increases the acid buffering capacity of the soil. It is expected that the carbonates will neutralize the H^+ ions generated at the anode which suppresses the development and migration of acidic pH front near the anode. Results obtained for electro kinetic remediation of glacial till by Reddy and his coworkers [36] have corroborated this finding. Bipolar effect was also investigated for EK-GAC-2, but pH gradient is not observed; hence, bipolar effect is not present.

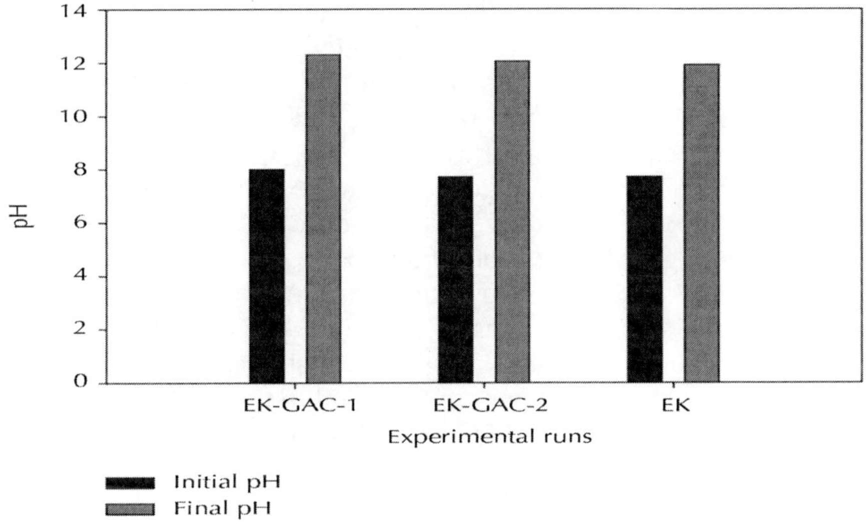

Figure 5: pH variation.

Soil Moisture Content, Organic Matter, and Electrical Conductivity

Soil moisture content enhances dissolved contaminant transport by ionic migration and electroosmosis and hence affects removal efficiency. In the present study, the GAC chamber was initially saturated with water, while the spiked soil specimen was kept at an initial moisture content of 33% in each case. This value increased to 52.6, 38.46, and 35.11% for EK-GAC-1, EK-GAC-2, and EK, respectively at the end of the 21-day period. Absence of GAC chambers in the EK run may be responsible for its lowest moisture content at the end of the run. Soil organic matter plays an important role in the adsorption of heavy metal ions even in soils where its value is very low [39]. This is because, SOM possesses very high specific surface area and cation exchange capacity (CEC) which may range between 150 and 300 cmol/kg [33]. The majority of a surface soil's CEC is in fact attributable to its soil organic matter. The initial SOM for the spiked chambers of EK-GAC-1, EK-GAC-2, and EK was 8.22, 6.38, and 6.38%, respectively. At the end of the experiments, these initial values decreased for EK-GAC-1 and EK-GAC-2 and increased slightly for EK.

Soil electrical conductivity (EC) varies with the amount of moisture held by soil particles. Electrical conductivity of clay typically lies between 0.01 and 1 dS/m. Abrol et al. [6] and Sparks [39] have classified soils whose EC and pH are greater than 4 dS/m and 8.2 (at 25°C) to be saline-sodic. The EC of the pristine clay sample is 8.62 dS/m which indicates that it has an excess of dissolved salts which makes it to be classified as saline-sodic soil with exchangeable sodium percentage (ESP) of more than 15. Upon spiking the soil, the EC jumped to 47.3 dS/m due to an increase in the dissolved ions and decreased at the end of the runs to 31.7 and 43.2 dS/m for EK-GAC-2 and EK, respectively (Figure 6). Higher reduction in EC of EK-GAC-2 may be explained by the higher contaminant removal efficiency. In the case of EK-GAC-1, the spiked EC was 24.56 dS/m which increased to 38.3 dS/m at the end of the test. This increase may be attributed to the higher voltage gradient (1 V/cm) used in this test which sped the rate of the electrochemical decomposition of water (at the electrodes) and degradation of the processing fluids.

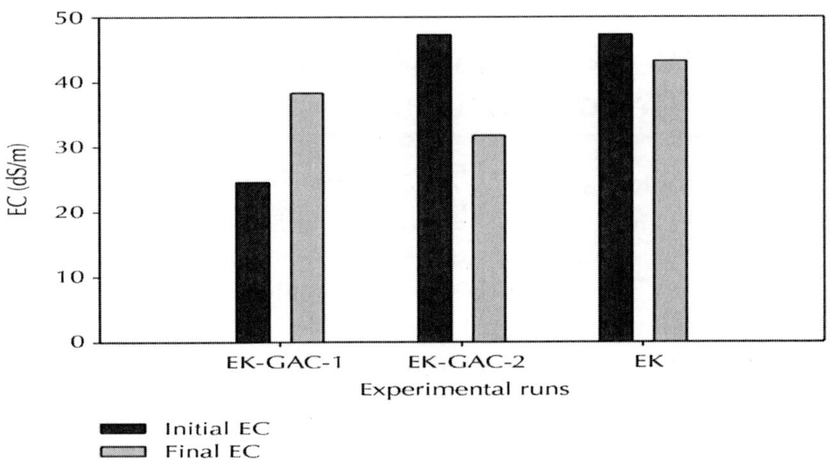

Figure 6: Variation of electrical conductivity.

Variations of Current, Temperature, and Cumulative Electroosmotic Flow

The average electric current recorded for EK-GAC-1, EK-GAC-2, and EK is 0.88, 0.61, and 0.71 A, respectively. Maximum current of 2.8 A was recorded by EK-GAC-1 test and may be attributed to its higher voltage gradient which facilitates faster ionic movement in pore fluid. Dynamic changes in the solution chemistry may be responsible for the observed fluctuating current trend observed in all the tests. Maturi and Reddy (2008) [12] observed somewhat similar current fluctuation. The current values recorded in the tests are 2-3 orders of magnitude higher than those obtained in similar studies employing the Lasagna process and electrokinetics only. This unique and important observation may be explained by the sodicity of the investigated soil which provides large amount of dissolved ions in the pore fluid for effective current conduction. High current flow through the soil may have significant impact on the soil temperature, electroosmotic flow rate, electrode and processing fluid deterioration, removal efficiency, and energy consumption. The average temperature recorded for EK-GAC-2 and EK is 28.5 and 30°C, respectively. For the same tests, the maximum temperature is 34.6 and 40.5°C, respectively. High temperature will reduce the soil moisture content due to pore fluid evaporation and subsequent reduction in current and electroosmotic flow.

Though thermal effects due to temperature rise have not been reported to be significant in bench-scale studies [16], our finding from this study reveals that increasing the voltage gradient more than 1 V/cm leads to considerable rise in the soil temperature which may not be neglected for practical purposes and modeling studies. As such, the general notion of using 1 V/cm for most bench-scale studies needs to be investigated if the soil possesses some properties that were not studied before. Electroosmotic flow is maintained throughout the duration of the tests, EK-GAC-2 and EK tests maintain an average pore volume of 0.75 and 0.66, respectively. Total electroosmotic volumes are 1388 and 1214 mL which are translated into total pore volumes flushed to be 17.35 and 15.17 for EK-GAC-2 and EK tests, respectively. Expectedly, the maximum temperature and electroosmotic flow recorded coincided with the period in which maximum current was recorded. Electroosmotic flow is not influenced by hydraulic gradient in this study as it occurs even under negative hydraulic gradient. Higher electroosmotic flow is expected to occur in the test with higher voltage gradient. It is obvious that the soil zeta potential is not reversed in this study which could reverse the electroosmotic flow. This is because it remains unidirectional throughout the test period.

Contaminant Removal Efficiency

After the operational period of 21 days, significant removal is observed for most of the contaminants in all the tests as presented in Figure 7. Highest and lowest percent removal is observed in phenol and Zn ions, respectively. Only phenol achieved 100% removal possible because it is miscible with water and behaves in the same way as other cationic species [40]. Acar and his coworkers [41] achieved similar percent removal after two pore volumes were flushed. Lukman et al. [24] observed that Zn is the least selective by this soil in competitive aqueous medium, most especially in the alkaline region. From Zn speciation using its Pourbaix diagram [42], it may be said that Zn precipitates as zinc hydroxide at the initial soil pH despite its existence in the form of hydroxocomplexes at pH>11. Calcareous soils, similar to the one studied here, have been found to perform relatively poor for zinc removal [43].

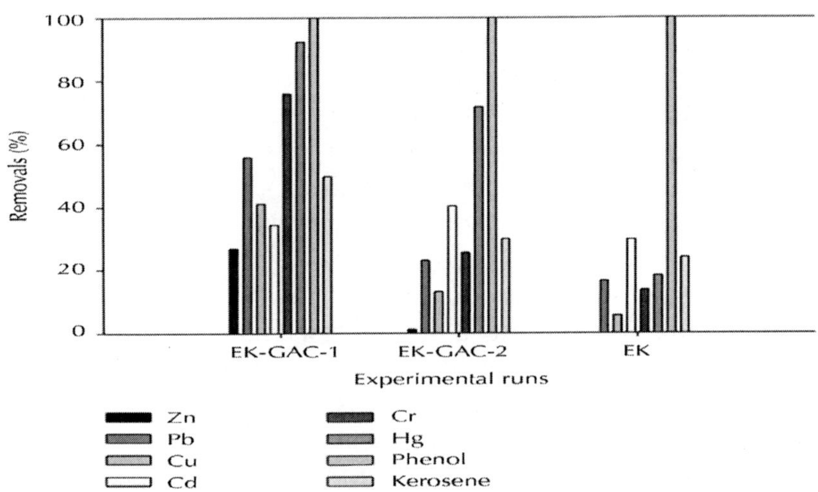

Figure 7: Comparison of the contaminant removal efficiencies for all the tests.

Among the trace elements studied in EK-GAC-1 and EK-GAC-2 tests, Hg removal was highest (92.49%). This may be attributed to the presence of excess Cl^- under aerobic conditions and subsequent formation and migration of the mercury complex $HgCl_4^{\ 2-}$ according to the following reaction [44]:

$$O_2 + 2Hg + 8Cl^- + 2H_2O \rightarrow 2HgCl_4^{2-} + 4O \quad (1)$$

Visual MINTEQ 3.0 was employed to model the Hg speciation before and after the first experimental run (EK-GAC- 1) using its dissolved concentration, pH, temperature, and ionic strength. It was assumed that all dissolved Hg species are removed at the end of the experimental run (i.e. 3 weeks). Table 2 provides the hydroxocomplexes present. The species having the highest initial % of total Hg concentration and removal efficiency is $Hg(OH)_2$ (aq) which suggests that precipitated mercury hydroxides can redissolve at high alkaline pH values encountered in this study. The formation of complexes with OH^- anions that have increased pore fluid solubility may be attributed to the redissolution of $Hg(OH)_2$ (aq) and its subsequent high removal efficiency. Each of the species has been removed by the process, with the dominant removal of $Hg(OH)_2$ (aq) corresponding to the actual removal efficiency of Hg during this experimental run (Figure 7).

Table 2: Modeling of Hg speciation using Visual MINTEQ 3.0 for EK-GAC-1

Initial Hg species	Concentration, mol/L	Final Hg species	Concentration, mol/L	Removed
Hg(OH)2 (aq)	4.30e — 04	Hg(OH)2 (aq)	3.98e — 04	92.49
Hg÷2	9.04e — 14	Hg÷2	1.73e — 22	1.91e — 07
Hg20H + 3	4.23e — 22	Hg20H÷3	4.03e — 35	9.53e — 12
Hg3(OH)3 + 3	1.79e — 22	Hg3(OH)3+3	3.77e — 35	2.1 le — 11
Hg0Ht	2.82e — 09	HgOlit	1.11e— 13	3.93e — 03

Figure 8 depicts how precipitated metal hydroxides can redissolve at high pH values. This occurs due to the formation of complexes with OH^- anions. These complexes are negatively charged and have increased pore fluid solubility. Consequently, the removal efficiency of these heavy metals is enhanced even under high alkaline condition prevailing in the present study.

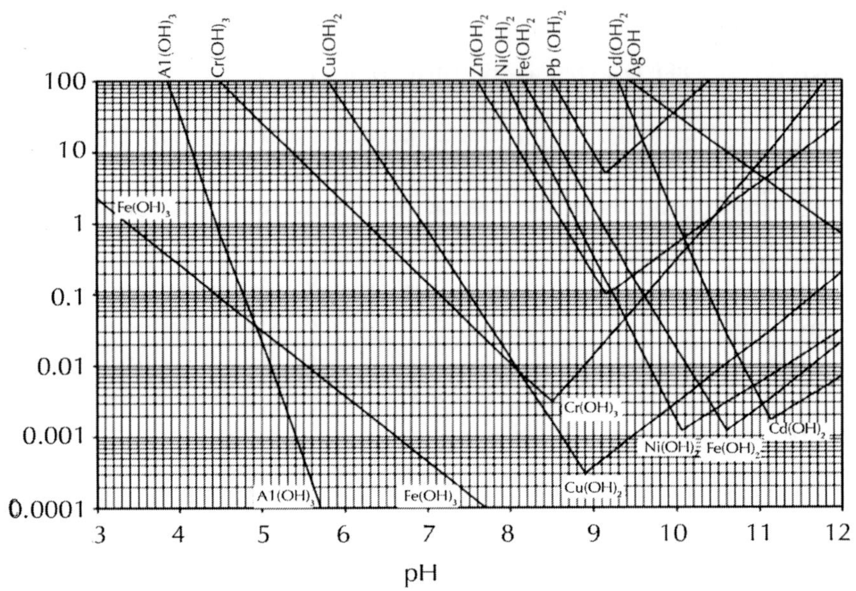

Figure 8: Variation of theoretical solubilities of some heavy metal hydroxides with pH [45].

Higher electroosmotic flow observed in EK-GAC-1 and EK-GAC-2 tests may be responsible for higher removal of kerosene in these tests than EK test (Figure 7). Generally, introducing GAC chambers in EK-GAC-2 test leads to higher percent removal of all contaminants than the case without the GAC chambers (EK test). In addition, higher voltage gradient produced higher percent removal (EK-GAC-1 and EK-GAC-2 tests).

Conditioning and Energy Consumption

Due to the rapid electrochemical decomposition of water at the electrodes and subsequent generation of H^+ and OH^- ions at the anode and cathode, respectively, there was a need to condition the anode and cathode chambers with H^+ and OH^- neutralizing chemicals. 2 N NaOH and 1 N HNO_3 were used as the anolyte and catholyte, respectively. Automatic processing fluid recycling was intended, but due to the soil sodicity which led to the passage of high current in the soil, periodic monitoring of the processing fluids pH was necessary. For the EK-

GAC-1 test, the catholyte becomes completely basic (pH $\approx$13) after 6 hr, while the anolyte lasted up to 12–18 hr before becoming completely acidic (pH $\approx$0.5). For the EK-GAC-2 and EK tests, the catholyte may last up to 12 hr before it needed replacement.

Higher current flow in EK test led to higher rate of processing fluids deterioration and higher energy consumption. The total energy consumed per $m^3$of soil treated during the 21- day period is estimated at 8.2, 3.4, and 4 kWhr/ $m^3$for EK-GAC-1, EK-GAC-2, and EK tests, respectively.

CONCLUSIONS

The potential of coupling electrokinetics and adsorption using locally produced granular activated carbon from date palm pits for the remediation of natural sodic soil was investigated. The soil was spiked with kerosene, phenol, Cr, Cd, Cu, Zn, Pb, and Hg at given concentrations and three tests (EK-GAC-1, EK-GAC-2, and EK) were run for a period of 21 days. Application of voltage gradient of 1 V/cm to the spiked soil led to high current generation, high electroosmotic flow, high rate of deterioration of processing fluids and anode electrodes, high percent removal, and high energy consumption. Phenol and Zn were found to have the highest and lowest removal efficiency. For the 21 day period of continuous electrokinetics-adsorption experimental run, efficiency for the removal of Zn, Pb, Cu, Cd, Cr, Hg, phenol, and kerosene was found to reach 26.8, 55.8, 41.0, 34.4, 75.9, 92.49, 100.0, and 49.8%, respectively. Despite the high selectivity of trivalent Cr exhibited by the clay soil, the alkaline pH maintained for most of the experimental duration has led to the formation of hydroxocomplexes which were removed due to electromigration. Hence, high percentage removal of the trivalent Cr was recorded. The results obtained suggest that integrating adsorption into electrokinetic technology is a promising solution for removal of contaminant mixture from sodic soils. It is suggested that different types of electrodes should be investigated (for this type of soil) together with the operating parameters (such as polarity reversal rate, pulse and continuous current application) affecting percent removal for simultaneous optimization of the Lasagna process.

ACKNOWLEDGMENTS

The authors would like to acknowledge the support provided by King Abdul-Aziz City for Science and Technology (KACST) through the Science & Technology Unit at King Fahd University of Petroleum & Minerals (KFUPM) for funding this work through Project no. 11-Env1669-04, as part of the National Science, Technology and Innovation Plan.

REFERENCES

1. S. V. Ho, P. W. Sheridan, C. J. Athmer et al., "Integrated in situ soil remediation technology: the Lasagna process," Environmental Science and Technology, vol. 29, no. 10, pp. 2528–2534, 1995.
2. K. Reddy, C. Cameselle, and P. Ala, "Integrated electrokinetic-soil flushing to remove mixed organic and metal contaminants," Journal of Applied Electrochemistry, vol. 40, no. 6, pp. 1269–1279, 2010.
3. J. I. Chang and C.-C. Lin, "A study of storage tank accidents," Journal of Loss Prevention in the Process Industries, vol. 19, no. 1, pp. 51–59, 2006.
4. K. Reddy, "Special issue on contaminant nixtures: fate, transport, and remediation," Journal of Hazardous, Toxic, and Radioactive Waste, vol. 15, no. 3, pp. 128–129, 2011.
5. V. P. Evangelou, Environmental Soil and Water Chemistry: Principles and Applications, John Wiley & Sons, Chichester, UK, 1998.
6. I. P. Abrol, J. S. P. Yadav, and F. I. Massoud, Salt-Affected Soils and Their Management, FAO Soils Bulletin no. 39, Food and Agriculture Organization of the United Nations, Rome, Italy, 1988.
7. C. J. Athmer and S. V. Ho, "Field studies: organic-contaminated soil remediation with Lasagna technology," in Electrochemical Remediation Technologies for Polluted Soils, Sediments and Groundwater, K. R. Reddy and C. Cameselle, Eds., pp. 625–646, John Wiley & Sons, Hoboken, NJ, USA, 2009.

8. K. Maturi and K. R. Reddy, "Simultaneous removal of organic compounds and heavy metals from soils by electrokinetic remediation with a modified cyclodextrin," Chemosphere, vol. 63, no. 6, pp. 1022–1031, 2006.
9. M. Elektorowicz, "Electrokinetic remediation of mixed metals and organic contaminants," inElectrochemical Remediation Technologies For Polluted Soils, Sediments and Groundwater, pp. 315–331, John Wiley & Sons, Hoboken, NJ, USA, 2009.
10. K. R. Reddy, "Integrated electrokinetic remediation technologies: opportunities and challenges," inProceedings of the 6th Symposium on Electrokinetic Remediation (EREM '07), Vigo, Spain, June 2007.
11. K. Reddy, K. Maturi, and C. Cameselle, "Sequential electrokinetic remediation of mixed contaminants in low permeability soils," Journal of Environmental Engineering, vol. 135, no. 10, pp. 989–998, 2009.
12. K. Maturi and K. R. Reddy, "Cosolvent-enhanced desorption and transport of heavy metals and organic contaminants in soils during electrokinetic remediation," Water, Air, and Soil Pollution, vol. 189, no. 1–4, pp. 199–211, 2008.
13. K. R. Reddy, P. R. Ala, S. Sharma, and S. N. Kumar, "Enhanced electrokinetic remediation of contaminated manufactured gas plant soil," Engineering Geology, vol. 85, no. 1-2, pp. 132–146, 2006.
14. K. R. Reddy and M. R. Karri, "Effect of oxidant dosage on integrated electrochemical remediation of contaminant mixtures in soils," Journal of Environmental Science and Health A, vol. 43, no. 8, pp. 881–893, 2008.
15. D. Huang, Q. Xu, J. Cheng, X. Lu, and H. Zhang, "Electrokinetic remediation and its combined technologies for removal of organic pollutants from contaminated soils," International Journal of Electrochemical Science, vol. 7, pp. 4528–4544, 2012.
16. S. V. Ho, C. J. Athmer, P. W. Sheridan, and A. P. Shapiro, "Scale-up aspects of the Lasagna process for in situ soil decontamination," Journal of Hazardous Materials, vol. 55, no. 1–3, pp. 39–60, 1997.

17. S.V. Ho, C. Athmer, P. W. Sheridan et al., "The Lasagna technology for in situ soil remediation. 2. Large field test," Environmental Science and Technology, vol. 33, no. 7, pp. 1092–1099, 1999.
18. S. V. Ho, B. M. Hughes, P. H. Brodsky, J. S. Merz, and L. P. Egley, "Advancing the use of an innovative cleanup technology: case study of Lasagna," Remediation Journal, vol. 9, pp. 103–116, 1999.
19. S.V. Ho, C. Athmer, P. W. Sheridan et al., "The Lasagna technology for in situ soil remediation. 1. Small field test," Environmental Science and Technology, vol. 33, no. 7, pp. 1086–1091, 1999.
20. B. D. Swift and J. J. Tarantino, "Application of the Lasagna soil remediation technology at the DOE paducah gaseous diffusion plant," in Proceedings of the Waste Management Symposium (WM '03), Tucson, Ariz, USA, February 2003.
21. J. W. Ma, F. Y. Wang, Z. H. Huang, and H. Wang, "Simultaneous removal of 2,4-dichlorophenol and Cd from soils by electrokinetic remediation combined with activated bamboo charcoal," Journal of Hazardous Materials, vol. 176, no. 1–3, pp. 715–720, 2010.
22. J. W. Ma, H. Wang, and R. R. Li, "Removal of cadmium in kaolin by electrokinetics-bamboo charcoal adsorption," Environmental Chemistry, vol. 26, pp. 634–637, 2007 (Chinese).
23. J. W. Ma, H. Wang, and Q. Luo, "Movement-adsorption and its mechanism of Cd in soil under combining effects of electrokinetics and a new type of bamboo charcoal," Chinese Journal of Environmental Science, vol. 28, no. 8, pp. 1829–1834, 2007 (Chinese).
24. S. Lukman, M. H. Essa, N. D. Mu'azu, A. Bukhari, and C. Basheer, "Adsorption and desorption of heavy metals onto natural clay material: influence of initial pH," Journal of Environmental Science and Technology, vol. 6, pp. 1–15, 2013.
25. ASTM, Annual Book of Standards: Soil and Rock 04.08, American Society for Testing & Materials, West Conshohocken, Pa, USA, 2004.
26. M. H. Essa and M. A. Al-Zahrani, "Date pits as potential raw material for the production of activated carbons," International Journal of Applied Environmental Sciences, vol. 4, no. 1, pp. 47–58, 2009.

27. M. S. Vohra, M. A. Al-Zahrani, M. H. Essa, and M. M. Rahman, "Synthetic industrial wastewater treatment using GAC produced from date palm tree branches," in Proceedings of the 8th International Conference and Exhibition on Chemistry in Industry (CHEMINDIX '10), A. S. Al-Dhuwaihi, Ed., Gulf International Convention Center, Gulf Hotel, Manama, Bahrain, 2010.
28. EPA, Method 3050B—Acid Digestion of Sediments, Sludges, and Soils, United States Environmental Protection Agency, Washington, DC, USA, 1996.
29. USEPA, Method 7000B—Flame Atomic Absorption Spectrophotometry, United States Environmental Protection Agency, Washington, DC, USA, 2007.
30. USEPA, Method 7473—Mercury in Solids and Solutions by Thermal Decomposition, Amalgamation, and Atomic Absorption Spectrophotometry, United States Environmental Protection Agency, Washington, DC, USA, 2007.
31. USEPA, Method 3545c: Pressurized Fluid Extraction (PFE), United States Environmental Protection Agency, Washington, DC, USA, 1996.
32. USEPA, Method 8270D—Semivolatile Organic Compounds by Gas Chromatography/Mass Spectrometry (GC/MS), United States Environmental Protection Agency, Washington, DC, USA, 2007.
33. F. J. Stevenson, Humus Chemistry: Genesis, Composition, Reactions, John Wiley & Sons, New York, NY, USA, 1982.
34. P. Srivastava, B. Singh, and M. Angove, "Competitive adsorption behavior of heavy metals on kaolinite,"Journal of Colloid and Interface Science, vol. 290, no. 1, pp. 28–38, 2005.
35. R. N. Yong, M. Nakano, and R. Pusch, Environmental Soil Properties and Behaviour, CRC Press, Boca Raton, Fla, USA, 2012.
36. K. R. Reddy, U. S. Parupudi, S. N. Devulapalli, and C. Y. Xu, "Effects of soil composition on the removal of chromium by electrokinetics," Journal of Hazardous Materials, vol. 55, no. 1–3, pp. 135–158, 1997.
37. S. Chinthamreddy and K. R. Reddy, "Oxidation and mobility of trivalent chromium in manganese-enriched clays during electrokinetic remediation," Journal of Soil Contamination, vol. 8, no. 2, pp. 197–216, 1999.

38. K. R. Reddy and S. Chinthamreddy, "Effects of initial form of chromium on electrokinetic remediation in clays," Advances in Environmental Research, vol. 7, no. 2, pp. 353–365, 2003.
39. D. L. Sparks, Environmental Soil Chemistry, Academic Press, San Deigo, Calif, USA, 2nd edition, 2003.
40. Y. B. Acar, R. Gale, A. N. Alshawabkeh et al., "Electrokinetic remediation: basics and technology status,"Journal of Hazardous Materials, vol. 40, no. 2, pp. 117–137, 1995.
41. Y. B. Acar, H. Li, and R. Gale, "Phenol removal from kaolinite by electrokinetics," Journal of Geotechnical Engineering, vol. 118, no. 11, pp. 1837–1852, 1992.
42. B. Beverskog and I. Puigdomenech, "Revised pourbaix diagrams for zinc at 25–300°C," Corrosion Science, vol. 39, no. 1, pp. 107–114, 1997.
43. L. M. Ottosen, H. K. Hansen, and P. E. Jensen, "Electrokinetic removal of heavy metals," inElectrochemical Remediation Technologies for Polluted Soils, Sediments and Groundwater, pp. 95–126, John Wiley & Sons, Hoboken, NJ, USA, 2009.
44. H. K. Hansen, L. M. Ottosen, B. K. Kliem, and A. Villumsen, "Electrodialytic remediation of soils polluted with Cu, Cr, Hg, Pb and Zn," Journal of Chemical Technology & Biotechnology, vol. 70, pp. 67–73, 1997.
45. E. R. Weiner, Applications of Environmental Aquatic Chemistry: A Practical Guide, Taylor & Francis Group, New York, NY, USA, 2008.

Chapter 6

A Modelling Approach to Evaluate the Long-Term Effect of Soil Texture on Spring Wheat Productivity under a Rain-Fed Condition

Yong He, Lingling Hou, Hong Wang, Kelin Hu, and Brian McConkey

[1]National Research Council, Saskatoon, Saskatchewan S7N0W9, Canada,

[2]Institute of Geographic Sciences and Natural Resources Research, Chinese Academy of Sciences, Beijing, 100101, P.R. China,

[3]Semiarid Prairie Agricultural Research Centre, Agriculture and Agri-Food Canada, Swift Current, Saskatchewan S9H3X2, Canada,

[4]College of Resources and Environmental Sciences, China Agricultural University, Beijing, 100193, P.R. China.

ABSTRACT

Soil surface texture is an important environmental factor that influences crop productivity because of its direct effect on soil water and complex interactions with other environmental factors. Using 30-year data, an agricultural system model (DSSAT-CERES-Wheat) was calibrated and validated. After validation, the modelled yield and water use (WU) of spring wheat (*Triticum aestivum L.*) from two soil textures (silt loam and clay) under rain-fed condition were analyzed. Regression analysis showed that wheat grown in silt loam soil is more sensitive to WU than wheat grown in clay soil, indicating that the wheat grown in clay soil has higher drought tolerance than that grown in silt loam. Yield variation can be explained by WU other than by precipitation use (PU). These results demonstrated that the DSSAT-CERES-Wheat model can be used to evaluate the WU of different soil textures and assess the feasibility of wheat production under various conditions. These outcomes can improve our understanding of the long-term effect of soil texture on

spring wheat productivity in rain-fed condition.Subject terms:

INTRODUCTION

Understanding of the variable response of crop to the environment is vital to crop management in agricultural systems. Available growing season precipitation is the most influential factor[1], to the productivity of spring wheat (*Triticum aestivum* L.) in the semiarid Canadian prairies. Productivity is also influenced by other environmental and management factors, including temperature, fertilizer rate, crop rotation, tillage, and cultivar[2, 3, 4].

The soil texture is another factor that influences wheat productivity because of its significant influence on ecological and hydrological processes, such as water retention, ion exchange and nutrient cycling[5, 6, 7]. The study of soil texture effects under field conditions is challenging using short-term data and classical statistics because of confounding effects of factors in the field, such as precipitation, that interact with soil texture effects. For example, conflicting effects of texture were found by Turner *et al.* (2004)[8] under a rain-fed growing condition. They found that the proportion of soil clay content has either a positive or negative effect on the water use efficiency (WUE) of wheat. This confusing result could be caused by many confounding factors acting together to reverse the soil texture effect on WUE depending on exact growing conditions.

Agricultural system simulation models that capture current knowledge of agricultural science have the potential to elucidate the effect of soil textures on crop productivity and water use (WU). Agricultural system simulation models, such as DSSAT (Decision Support System for Agrotechnology Transfer) Crop System Models, are often used as tools to estimate the environmental impacts on field crop production. A number of previous studies have used DSSAT-CERES to discern various influences on crop productivity, such as temperature[9], irrigation[10, 11], nitrogen cycling[12], tillage[13], and crop rotation[14]. However, there are few studies in the literature[15]that investigate the effect of soil texture, another important factor influencing spring wheat productivity. Our previous study[16] simulated the crop productivity in different soil textures but predominantly focused on climate effect on seeding dates. Therefore, there is a need to evaluate the capability of DSSAT-CERES-Wheat model to simulate the effect of soil texture on crop productivity. It is also essential to demonstrate if DSSAT-CERES-Wheat model can be used to estimate the WU difference caused by soil texture.

Our objectives were to: 1) calibrate the DSSAT-CERES-Wheat model for simulation of wheat productivity in two soil textures (silt loam and clay soil); 2) validate the calibrated model by comparing simulated and measured wheat productivity and soil moisture data; and 3) compare the estimated water use (WU) difference between these two soil textures from the simulation results.

METHODS

Study Sties and Experimental Set-Up

The study was conducted at two sites with contrasting soil texture that are located in the drier portion of the Canadian prairies[17]. The sites are about 30 km apart. These sites are located at southwest Saskatchewan, Canada on Brown chernozem soils[18]. The land was slightly sloping (0.5 to 2%). The annual average temperature and total precipitation near here is 6.9°C and 332 mm, respectively[16]. Precipitation during the summer and fall periods was usually in the form of rain, most of which evapotranspires, but some penetrates to deeper soil layers. During the winter, the precipitation was usually in the form of snow that lies on the frozen soil until March or April when it melts. Much of this snowmelt runs off and is lost to wheat production[19].

The data used in this study was from a long-term tillage experiment and only data for the conventional tillage treatment was used. The experiment established on a silt loam site (Swift Current, 50.288°N, 107.794°W) and a clay soil site (StewartValley, (50.596°N, 107.805°W). Randomized complete block designs were used, with four replications on the silt loam site, and three replications on the clay site.

Management (tillage, fertilizer rate, etc.) was as consistent as possible between the two sites. The same wheat varieties that have similar phenology and yield performance were sown at both locations each year. A tillage operation occurred just prior to seeding, usually a day or two before the seeding operation. Tillage depth was approximately 5 cm. In-crop herbicide application normally occurred in early to mid-June, around the 5 to 6 leaf stage of the crop. Details of the fertilizer management on these two sites have been reported previously[15, 19, 20]; consequently, only that pertinent to this study are presented. The fertilizer rates of the two sites were quite similar (Table 1). Grain yield and above ground biomass (Table 1) were determined at full ripeness. Total aboveground yield and grain yield from a 1 m^2 equivalent area were measured after air drying of the samples. A minimum of three samples were taken per plot for most measurements. Soil moisture was measured each year after harvest. Three 2.5-cm diameter cores were taken per plot with a truck-mounted hydraulic soil sampler. Gravimetric soil water content were determined for the 0–15, 15–30, 30–60, 60–90 and 90–120 cm depths and these values converted to volumetric basis using bulk densities.

Table 1: Growing season (May to August) precipitation (GSP, mm), nutrients applied, grain yield and biomass of wheat grown in silt loam and clay soil from 1982 to 2011

Year	GSP (mm)	Nutrients applied (NO3-N kg ha–1)		Grain yield (kg ha–1)		Biomass (kg ha–1)	
		Silt loam	Clay	Silt loam	Clay	Silt loam	Clay
Note: Values in parentheses are the standard deviation. '-' is missing value.							
1982	243	62	43	2839 (245)	1577 (63)	8443 (837)	3791 (216)
1983	186	55	55	2111 (56)	-	5355 (208)	-
1984	94	50	55	651 (154)	1166 (181)	1933 (353)	3062 (450)
1985	73	35	25	764 (146)	1156 (197)	1999 (158)	2552 (378)
1986	205	20	20	2873 (237)	-	6751 (551)	-
1987	129	45	45	1399 (161)	783 (124)	2796 (366)	1450 (251)
1988	143	38	5	532 (83)	-	1288 (176)	-
1989	210	19	23	2258 (157)	1663 (181)	6216 (201)	4881 (449)
1990	179	45	32	2485 (62)	2159 (150)	6153 (138)	5240 (486)
1991	302	55	55	2770 (275)	3052 (45)	7827 (627)	8669 (561)
1992	183	58	60	1866 (56)	2017 (28)	5240 (98)	4817 (54)
1993	175	60	55	2430 (97)	2783 (331)	5950 (208)	7167 (575)
1994	160	54	58	2384 (149)	2518 (340)	7553 (600)	6363 (859)

1995	189	61	64	2809 (313)	2499 (427)	8069 (706)	7459 (1264)
1996	168	53	60	2320 (175)	2276 (111)	6331 (695)	6263 (114)
1997	163	58	62	2775 (432)	2511 (148)	6689 (813)	6581 (394)
1998	166	64	63	1072 (171)	1337 (53)	4220 (427)	4041 (157)
1999	240	49	62	2416 (186)	3234 (256)	6395 (707)	7966 (817)
2000	245	65	66	2035 (54)	1962 (488)	6386 (310)	5415 (814)
2001	112	63	64	970 (255)	663 (101)	2072 (588)	1402 (216)
2002	237	32	25	2354 (388)	2100 (307)	5463 (806)	4643 (731)
2003	129	60	51	1675 (227)	2145 (152)	5048 (773)	4999 (394)
2004	211	46	46	3009 (83)	2690 (61)	7668 (147)	7305 (486)
2005	167	60	56	2273 (263)	2948 (205)	6566 (247)	8111 (422)
2006	269	58	60	1661 (196)	2037 (149)	4550 (377)	5620 (329)
2007	110	61	65	1264 (129)	2193 (138)	3482 (403)	5476 (596)
2008	253	40	55	2176 (260)	2970 (265)	4860 (736)	8428 (655)
2009	94	55	57	1374 (193)	2278 (215)	2782 (378)	4760 (454)
2010	286	49	54	2472 (282)	2932 (416)	6346 (780)	6997 (905)
2011	240	55	57	2880 (492)	2638 (156)	7478 (1241)	6387 (405)
Mean	185	51	49	2030	2159	5397	5352

Note: Values in parentheses are the standard deviation. '-' is missing value.

Model Modification of Seedling Emergence

The timing of seedling emergence greatly affects the growth and yield of wheat. The DSSAT-CSM is used worldwide for many different applications, but its simulation of the timing of seedling emergence of wheat is not satisfactory under certain circumstances[21]. The cereal crop seedling emergence is simulated by DSSAT-CSM model using thermal time and adjusted by a soil water factor, assuming that the adjusted thermal time is linearly related to the emergence process[22]. However, Jame and Cutforth (2004)[21] showed that the response of emergence to temperature is not linear. In order to improve the prediction of seedling emergence, we incorporated a newly developed non-linear model, the Beta model, into DSSAT-CSM. For a more detailed description, please refer to Wang *et al.* (2009)[23].

Model Input

The model requires daily weather data including maximum and minimum temperature and precipitation that were obtained from the Meteorological Service of Canada (MSC). The latter Daily solar radiation was calculated using the Mountain Climate Simulator[24]. The model also requires input parameters describing crop and soil characteristics. The soil parameters included lower soil water limit (LL, $cm^3\ cm^{-3}$), upper drainage limit (DUL, $cm^3\ cm^{-3}$), upper soil water limit (SAT, cm^3cm^{-3}), soil organic C content (wt.%), dry bulk density ($g\ cm^{-3}$), saturated hydraulic conductivity, soil pH, clay and silt content. The pH, clay and silt data were obtained from Saskatchewan Soil Survey reports. The soil hydraulic parameters (LL, DUL and SAT) were obtained by a soil texture based pedo-transfer function (the SPAW model) developed by Saxton and Rawls (2006)[25]. The root distribution parameter was estimated by a method provided by Gijsman *et al.* (2007)[26]. The data were used to construct the model soil profiles (Table 2).

Table 2: Selected soil physical properties at the slit loam and clay sites

Soil Textures	Soil profile depth (cm)	0–15	15–30	30–60	60–90	90–120
Silt loam	SLLL	0.12	0.11	0.10	0.09	0.09

	SDUL	0.29	0.29	0.27	0.27	0.25
	SSAT	0.47	0.47	0.47	0.47	0.47
	Root growth factor	1.00	0.95	0.82	0.47	0.26
	SSKS	1.42	2.76	4.32	3.04	3.04
	Clay content (%)	18.2	24.2	25.4	33.4	31.7
	Silt content (%)	50.4	49.4	33.4	25.9	27.9
Clay	SLLL	0.14	0.14	0.16	0.18	0.20
	SDUL	0.47	0.47	0.47	0.47	0.47
	SSAT	0.55	0.55	0.55	0.55	0.51
	Root growth factor	1.00	0.95	0.82	0.47	0.26
	SSKS	0.33	0.33	0.33	0.33	0.22
	Clay content (%)	59	59	59	59	69
	Silt content (%)	36	36	36	36	28

The soil moisture was simulated by the DSSAT model. Since historic soil water content before experiment initiation was unknown, the model was initialized to start three years prior to the period of analysis. A spin-up period of three years was found to be adequate for the soil water values to stabilize and not be affected by the initial input values[16]. Once the soil profile initial conditions were established, the simulations were run from 1982 to 2011.

Crop management data needed are seeding date, seeding depth, row spacing, and plant population.

Model Calibration

The method for calibrating parameters (soil, ecotype and cultivar) of the crop models is a step-by-step procedure following the approach recommended by Boote[27] of calibrating soil moisture first, followed by phenology, and then yield (biomass and grain yield). The initial soil hydraulic parameters were obtained by a soil texture based model

(SWAP[25]). Plant parameters (ecotype and cultivar) from a previous study[23] conducted in the same area and using same cultivar as our study were used as initial values for calibration. Model calibration used part (1982–1999) of the long-term data of the two soil textures (sites) and two (2004 and 2005) of the three years of data from a physiological study. A single set of ecotype and cultivar parameters was used since the cultivars grown from 1982 to 2011 have similar phenology and yield performance (Table 3). For the calibration, we fixed the maximum possible seed weight under non-stressed conditions at 33 g, the highest value reported by Wang *et al.* (2002)[28] using the same cultivar in our site. As little information on the other cultivar and ecotype parameters was available in the experiments or literature, they were mostly calibrated through trial and error to match simulations with measurements.

Table 3: Cultivar parameters and part ecotype parameters for wheat developed for simulation using the DSSAT-CERES-Wheat module

Parameters	Definitions	Calibrated values
P1V	Days at optimum vernalizing temperature required to complete vernalization	30
P1D	Percentage reduction in development rate in a photoperiod 10 hour shorter than the threshold relative to that at the threshold	55
P5	Grain filling (excluding lag) phase duration (°C.d)	490
G1	Kernel number per unit canopy weight at anthesis (#/g)	15
G2	Standard kernel size under optimum conditions (mg)	33
G3	Standard, non-stressed dry weight (total, including grain) of a single tiller at maturity (g)	0.8
PHINT	Interval between successive leaf tip appearances (°C.d)	75
HSTD	Standard canopy height (cm)	95

LAWR2	Lamina area to weight ratio, phase 2 (cm2/g)	140
LAWRS	Lamina area to weight ratio of standard first leaf (cm2/g)	200
P1	Duration of phase end juvenile to double ridges (PVTU)	240
P1DPE	Day length factor, pre emergence (#,0–1)	0.0
P2	Duration of phase double ridges to end leaf growth (TU)	180
P3	Duration of phase end leaf growth to end spike growth (TU)	160
P4	Duration of phase end spike growth to end grain fill lag (TU)	400
P4SGE	Stem growth end stage (Growth Stage)	4.45
TBGF	Temperature base, grain filling (°C)	0.0
PARUR	PAR conversion to dry material ratio, after last leaf stage (g/MJ)	2.5
PARUV	PAR conversion to dry material ratio, before last leaf stage (g/MJ)	2.2
WFGU	Water stress factor, growth, upper (fr)	1.0
WFPGU	Water factor, genotype sensitivity to stress when grain filling (0–1)	0.8
WFPU	Water stress factor, photosynthesis, upper (fr)	1.0

Statistics for Model Calibration and Validation

The inaccuracies in each simulation step will affect the accuracy of simulated results for the subsequent simulation step. These inaccuracies will add together across all simulation steps and affect final estimated results. Also, the inaccuracies of simulated soil moisture will affect the accuracy of the phenology simulation. Finally, inaccuracies in soil moisture and phenology will affect accuracy of simulated final yield. These accumulated effects on final results of phenological development, WU, and crop yield are complex so it is infeasible to determine how the inaccuracies at each step accumulate to affect these final results.

Instead, we determined the overall errors based on difference of each pairing of observed and simulated values. We evaluated the simulation results using: 1) RMSE (Eq.1); and 2) the index of agreement (d), which varies between 0 (poor model) and 1 (perfect model)[29] (Eq.2).

$$RMSE = \sqrt{\frac{1}{n}\sum_{i=1}^{n}(p_i - o_i)^2} \quad (1)$$

$$d = 1.0 - \frac{\sum_{i=1}^{n}(p_i - o_i)^2}{\sum_{i=1}^{n}\left(\left|p_i - o_{avg}\right| + \left|o_i - o_{avg}\right|\right)^2} \quad (2)$$

Where P_i is the *i*th simulated value, P_{avg} is the average of the simulated values, O_i is the *i*th measured value, O_{avg} is the average of the measured values, and *n* is the number of data pairs.

RESULTS

Model Calibration

From 1982 to 1999, wheat was grown in two soil textures. The grain yield and biomass collected in years 1983, 1986 and 1988 were missing, and thus excluded from the analysis. The two soil texture sites were simulated sequentially, with the simulation starting on 1 Jan. 1979 and ending on 31 Dec. 1999. Grain yield and biomass (at harvest) and soil moisture measurement (after harvest) were available for model calibration.

As shown in Table 4, soil moisture simulations in individual soil layers (0–60 cm) had RMSEs ranging from 0.020 to 0.055 (cm^3 cm^{-3}) and from 0.015 to 0.043 (cm^3 cm^{-3}) for the silt loam and clay sites, respectively. This indicates the soil moisture above 60 cm was well simulated. However, the soil moisture was not well simulated on both the silt loam and clay sites below 60 cm ($d < 0.593$). The variation in simulated soil moisture below 60 cm was smaller than those measured for both soil texture sites (Fig. 1 and 2). A plausible cause of the poor model performance is that the model assumed that half of the surplus water was held over to the next time step (i.e., the next month). This simple method of handling storage is not physically realistic and lowers the accuracy of the model simulations[28]. Thus, there is a need to improve model simulation of soil water in deeper soil layers. DSSAT simulates soil moisture more accurately during the growing season than before seeding or after harvest. This is not unexpected because DSSAT was primarily designed for simulating soil moisture-crop interactions[30]. The accuracy of soil moisture simulation is acceptable for estimating growing season WU.

Table 4: Mean measured values (calibration data set) and mean square error (RMSE), and d values in simulations of soil moisture in different depth of silt loam and clay soil

Soil profile depth	Silt loam (cm3 cm–3)				Clay (cm3 cm–3)			
(cm)	Mean	RMSE	d	Samples	Mean	RMSE	d	Samples
0–15	0.185	0.055	0.764	18	0.205	0.043	0.864	18
15–30	0.150	0.029	0.895	18	0.194	0.035	0.881	18
30–60	0.125	0.021	0.904	18	0.195	0.021	0.900	18
60–90	0.109	0.020	0.473	18	0.214	0.015	0.593	18

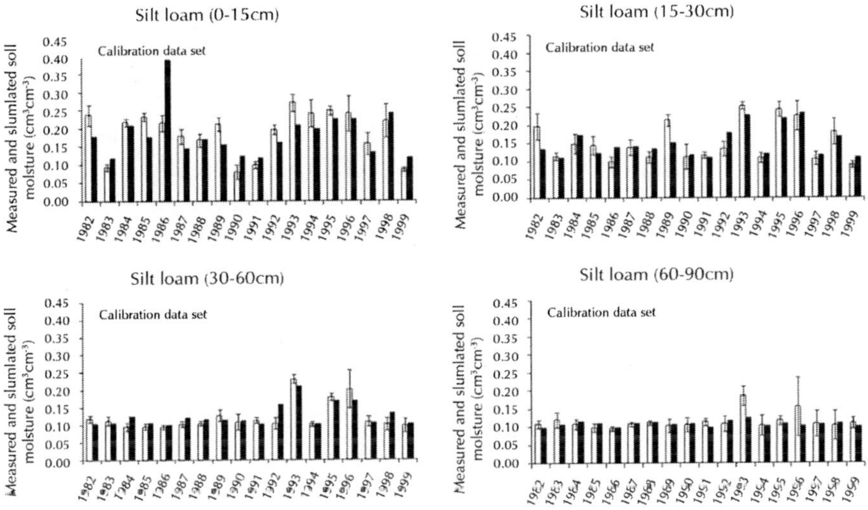

Figure 1: Comparison of measured (open bars with error bars) soil moisture after harvest and corresponding simulated (solid bars without error bars) soil moisture in layers at silt loam soil site.

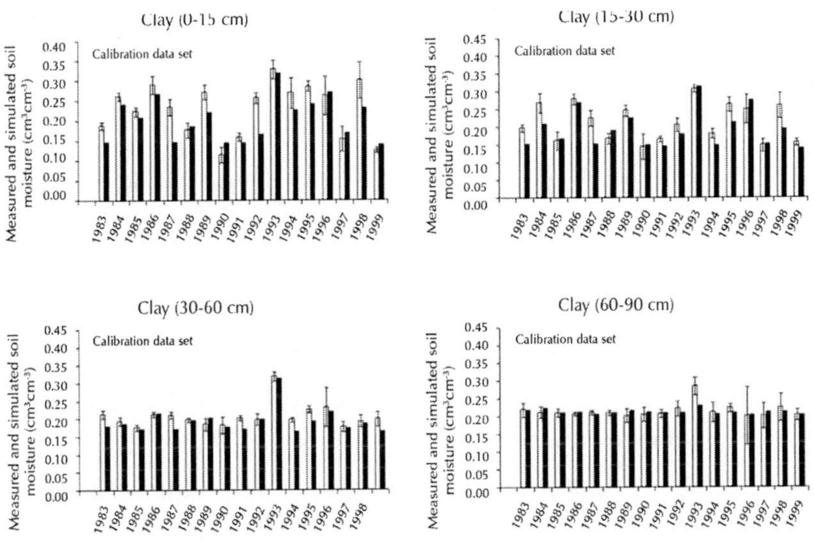

Figure 2: Comparison of measured (open bars with error bars) soil moisture after harvest and corresponding simulated (solid bars without error bars) soil moisture in layers at clay soil site.

The wheat phenology in the CSM- CERES-Wheat model is controlled by growth stages, which are in turn driven by growing degree-days (GDD)[31]. The cultivar coefficients defined in DSSAT are specific to both the crop cultivar and the local climate, and must therefore be individually calibrated[12]. The two years of simulation RMSEs of wheat phenology was less than 3.46 (days) and d values were higher than 0.99 (Table 5 and Fig. 3), indicating that the DSSAT model simulated the wheat growth very well.

Table 5: Calibration and validation of wheat phenology

	Year	RMSE	d	Samples
Calibration	2004	1.21	0.999	8
	2005	3.46	0.996	7
Validation	2006	3.152	0.997	6

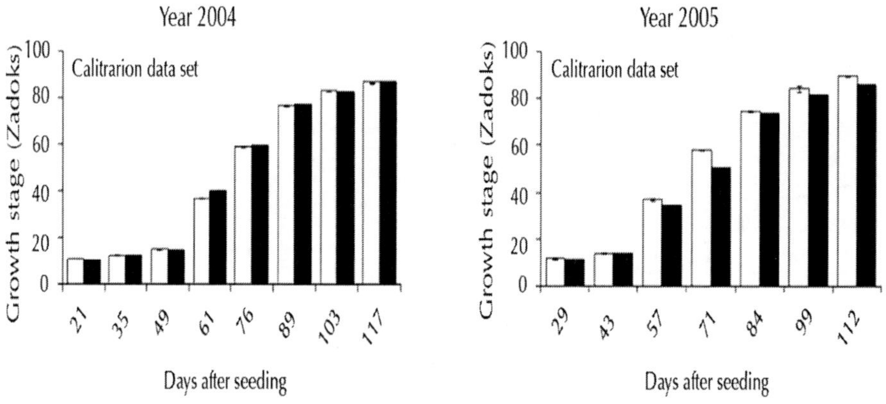

Figure 3: Comparison of measured (open bars with error bars) and simulated (solid bars without error bars) wheat phenology.

The simulated and measured grain yields for silt loam and clay from 1982 to 1999 are displayed inFig. 4 and 5. The annual fluctuation of grain yield and biomass were reasonably simulated for both soil texture sites. With the exception of the grain yield and biomass in 1995 in silt loam soil, all of the grain yield and biomass were simulated well with RMSEs of 777 (kg ha^{-1}), and d value of 0.808 for grain yield and an RMSE of 1871 (kg ha^{-1}), and d value of 0.836 for biomass (Table

6). The grain yield and biomass was also well simulated in clay soil with the exception of grain yield in 1993 and biomass in 1982. The deviations between simulated and measured grain yield and biomass in these individual years were not good for either soil texture sites, but the overall accuracy were on the same order as those obtained by others[32, 33]. The large differences between simulated and measured grain yield and biomass may be related to our assumptions regarding soil physical properties and cultivar characteristics. The imprecise simulation of grain yield and biomass in these individual years probably reflects factors and/or events that are not considered in the model, such as heavy rainfalls, high winds, weed competition, crop diseases, hail damage, etc[12].

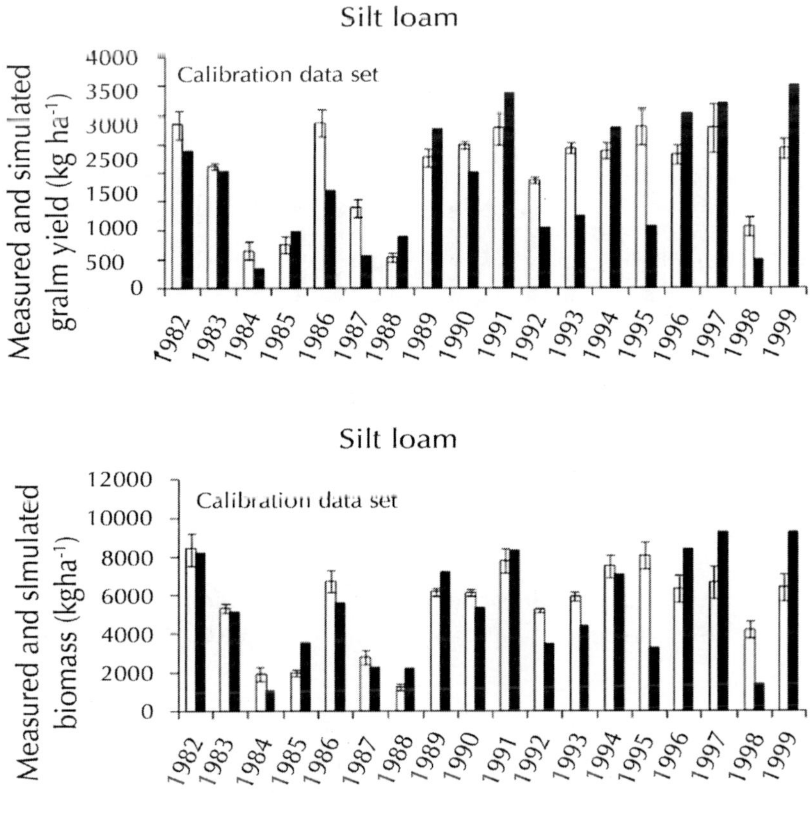

Figure 4: Comparison of measured (open bars with error bars) and simulated (solid bars without error bars) grain yield and biomass (kg ha^{-1}) at silt loam soil site.

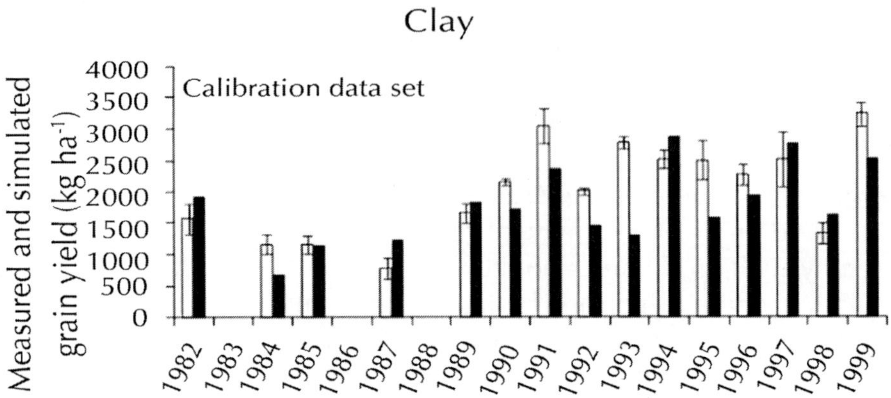

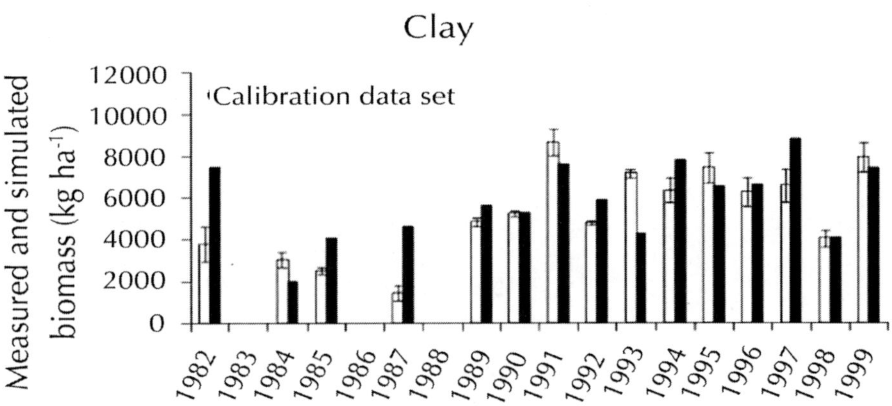

Figure 5: Comparison of measured (open bars with error bars) and simulated (solid bars without error bars) grain yield and biomass (kg ha^{-1}) at clay soil site.

Table 6: Mean measured values (calibration data set) and mean square error (RMSE), and d values in simulations of grain yield and biomass of silt loam and clay soil

	Grain yield (kg ha–1)				Biomass (kg ha–1)			
	Mean	RMSE	d	Samples	Mean	RMSE	d	Samples
Silt loam	2042	777	0.808	18	5512	1871	0.836	18
Clay	2049	599	0.791	15	5353	1785	0.766	15

Inspection of the data comparisons presented in Fig. 1 to 5 indicates that the model was adequately calibrated for soil moisture, wheat phenology, grain yield and biomass.

Model Validation

The results from the calibrated model were then compared with the remaining data, which were a continuation of the long-term study and the physiological study. The soil moisture simulations, in terms of RMSE in various soil layers of the silt loam and clay sites, ranged from 0.025 to 0.046 (cm^3cm^{-3}), and from 0.031 to 0.042 (cm^3 cm^{-3}), respectively (Table 7). Their corresponding d values were higher than 0.9 in all layers. Figure 6 and 7 shows that the simulated soil moisture in all layers are scattered more or less equally around the 1:1 line for all textures, indicating that there were no systematic deviations in the model simulation, and that the soil moisture was simulated reasonably well for both soil textures.

Table 7: Mean measured values (validation data set) and mean square error (RMSE), and d values in simulations of soil moisture in different depth of silt loam and clay soil

Soil profile depth	Silt loam (cm3 cm–3)				Clay (cm3 cm–3)			
(cm)	Mean	RMSE	d	Samples	Mean	RMSE	d	Samples
0–15	0.181	0.027	0.983	12	0.223	0.042	0.996	12
15–30	0.162	0.046	0.965	12	0.211	0.032	0.999	12
30–60	0.131	0.037	0.980	12	0.209	0.036	0.999	12
60–90	0.111	0.025	0.990	12	0.219	0.031	0.995	12

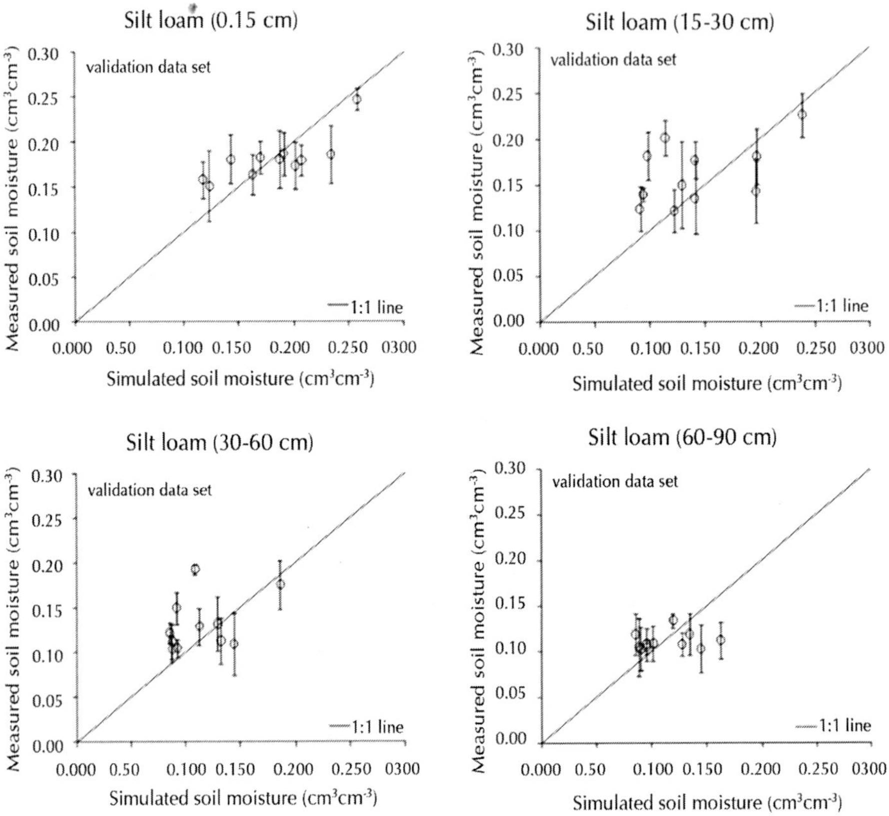

Figure 6: Comparison of measured (validation data set) vs. simulated soil moisture in layers at silt loam soil site.

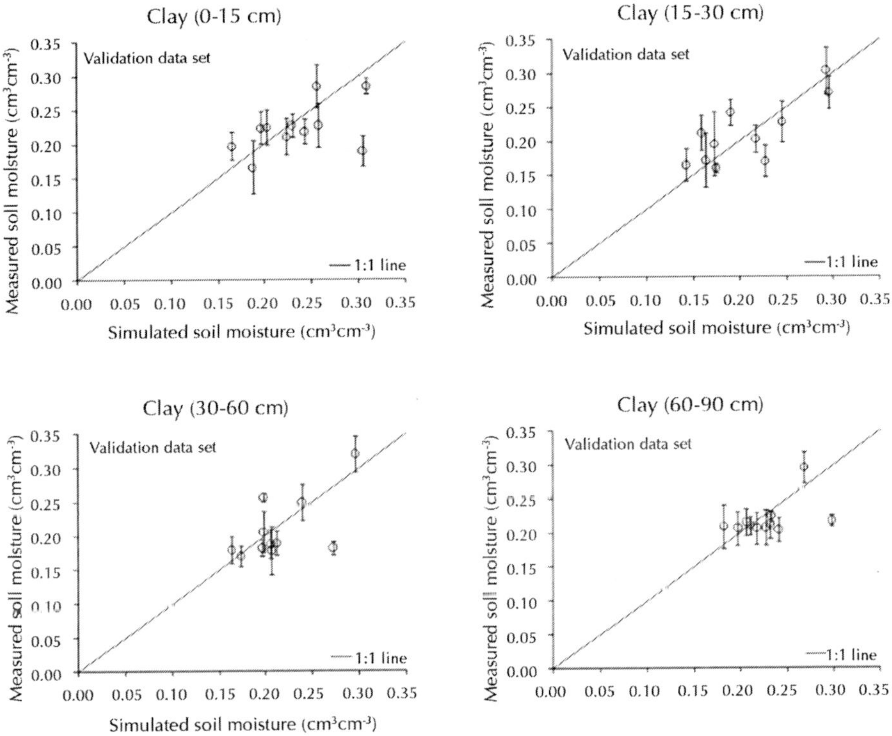

Figure 7: Comparison of measured (validation data set) vs. simulated soil moisture in layers at clay soil site.

Wheat phenology was fairly well simulated with both a low RMSE (3.15 days) and a high d value (0.997) in year 2006 (Table 5 and Fig. 8). Overall, the simulated mean and ranges of grain yields for the two soil texture sites corresponded closely to observed means and ranges (Fig. 9 and 10). The model also correctly simulated the biomass of the two soil texture sites with a relative low RMSE (<1688 kg ha^{-1}) and a high d value (>0.77) (Table 8), and equally distributed scattering around the 1:1 line for both soil textures (Fig. 9 and 10).

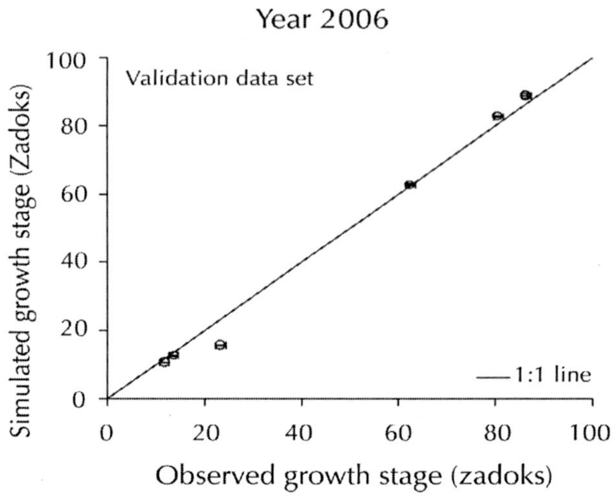

Figure 8: Comparison of measured (validation data set) vs. simulated wheat phenology in year 2006.

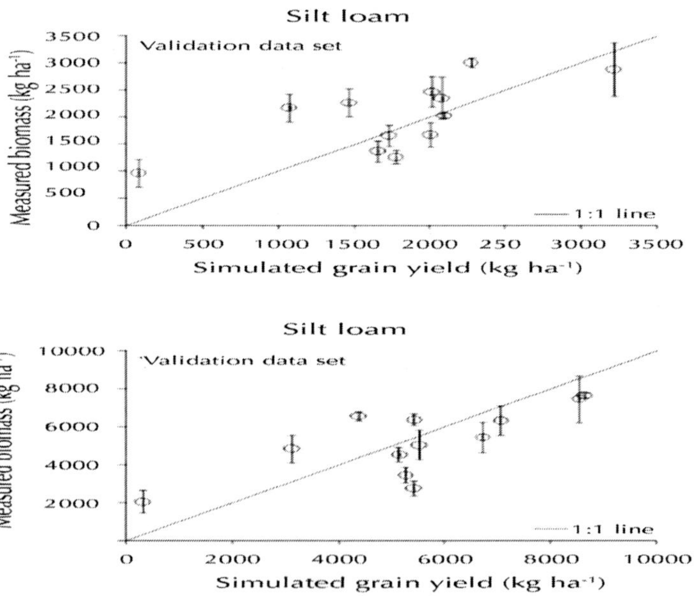

Figure 9: Comparison of measured (validation data set) vs. simulated grain yield and biomass at silt loam soil site.

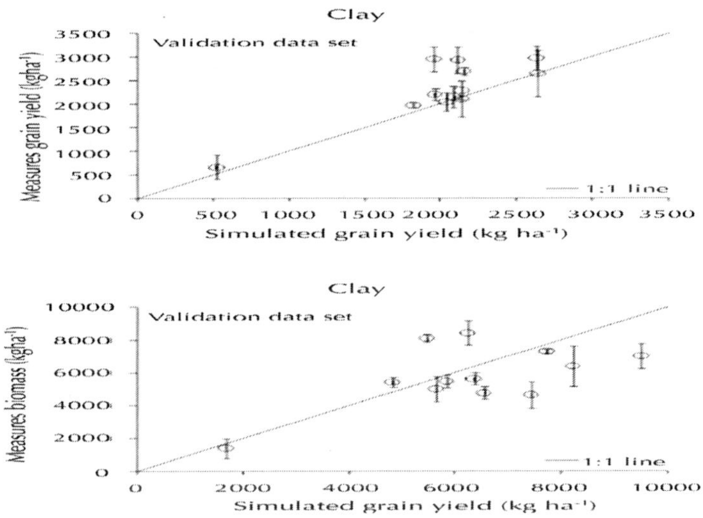

Figure 10: Comparison of measured (validation data set) vs. simulated grain yield and biomass at clay soil site.

Table 8: Mean measured values (validation data set) and mean square error (RMSE), and d values in simulations of grain yield and biomass of silt loam and clay soil

	Grain yield (kg ha–1)				Biomass (kg ha–1)			
	Mean	RMSE	d	Samples	Mean	RMSE	d	Sam-ples
Silt loam	2012	581	0.792	12	5225	1494	0.839	12
Clay	2296	426	0.868	12	5795	1688	0.770	12

Comparison of Water Use between Silt Loam and Clay Soil

After validation, the model was then used to compare the precipitation use (PU) and water use (WU) between the two soil textures based on the simulation outputs (soil moisture in profile, maturity, grain yield). The PU was defined as total precipitation received between seeding and maturity for each growing season. The WU was calculated as the difference of soil moisture (0 to 1.8 m depth) between seeding and maturity, plus PU[34]. As shown in Fig. 11, the WU for the two soil textures has a similar trend over the years due to the predominant common effect factor of precipitation amount during the growth period.

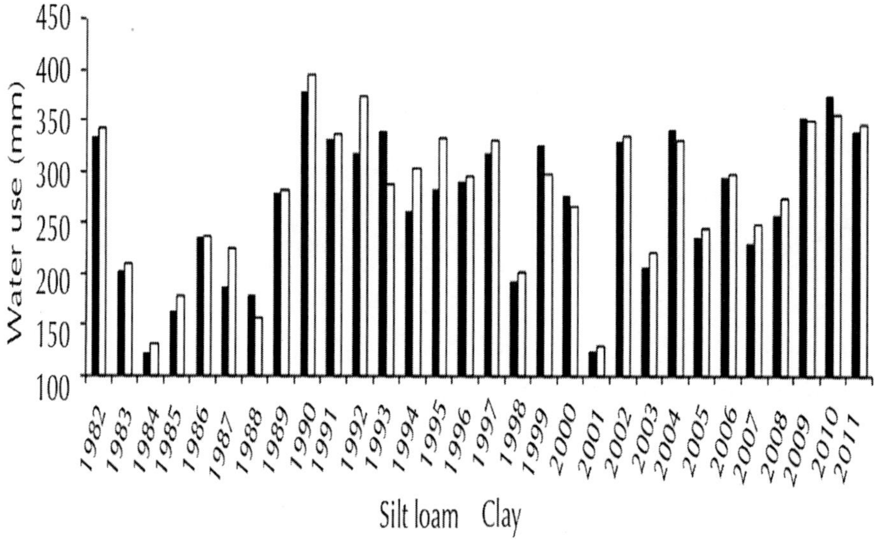

Figure 11: Comparison of simulated water use in silt loam and clay soil.

The difference in WU between the two soil textures are mainly affected by soil differences in water holding capacity. These differences ranged from 2 to 57 mm. A minor amount of the differences in WU would be due to the 1 to 7 day differences between the seeding dates at the two sites. A noticeable trend was found that the clay soil has higher WU than silt loam soil in most drought years (defined as those years, 1984, 1985, 1987, 2001, 2007 and 2009, that fell into the lowest quantile of growing season precipitation). This result explains why the clay soil out-yields the silt loam soil in most drought years (Table 1).

The relationship between grain yield and the WU was linear for both silt loam and clay soils (Fig. 12). The slopes for the two equations are 8.245 and 5.962, respectively, and are assumed to provide some measure of sensitivity of grain yield to WU. The larger slope for wheat grown on a silt loam soil shows it is more sensitive to WU than wheat grown on a clay soil. The R^2 of fitted equations of simulated

WU and simulated yield of silt loam was higher than that of clay soil, also indicating that silt loam is more sensitive than clay soil, which coincides with our previous statistical analysis[15]. The R^2 of fitted linear relationships between simulated WU and simulated yield were much higher than those between PU and simulated yield for both silt loam and clay soils. This demonstrates that, yield variation is much better be explained by WU than by PU. This result clearly shows that soil t

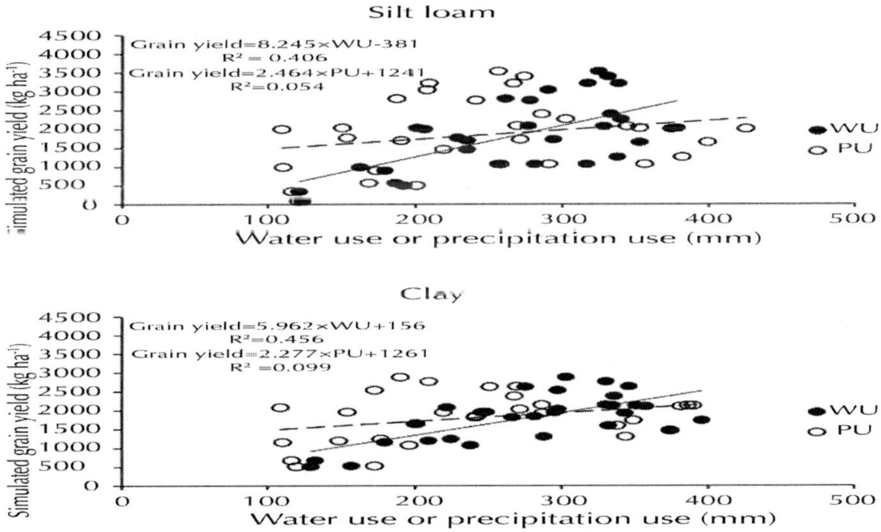

Figure 12: Liner relationships between water use (WU) or precipitation use (PU) and simulated grain yield in silt loam and clay soil over 30 years (1982–2011).

DISCUSSION

The objective of this study is to evaluate the effect of soil texture on wheat productivity in rain-fed condition. However, to understand the variation led by soil texture is challenging due to a lot of confounding factors. To achieve the goal of our objective, we adopted an agricultural system model (CSM-CERES-Wheat model). It is largely recognized that the model need to be well calibrated and validated before applied for account for the impacts[35, 36]. Therefore, a long-term (1982–2011) continuous wheat experiment in southwest Saskatchewan, on both silt loam and clay soils was used to calibrate and validate the model. This data set is ideal for investigation of soil textures' impact, which because other factors such as cultivar characteristics and field management are all same in these two sites every year. The two sites also have very similar weather condition since only 30 km away from each other. Therefore, the main contribution to yield difference should due to soil texture.

The model performance for si

mulating yields, although not better than for soil moisture and phenology, was comparable to that of other research with simulation models[31, 32]. The large differences between simulated and measured grain yield and biomass for some individual years may be related to inaccuracies in our assumptions regarding soil physical properties and sampling procedures. For example, grain yield simulation in 1993 at clay soil site deviated from the measured by 2819 kg ha^{-1} (error of −39%). During this crop season, the model was not able to simulate the soil water and nitrite dynamics well, leading to a water stress (0.27) during tillering stage and high nitrite stress (0.64) during the grain filling stage resulting in under prediction of biomass and grain yield (data not shown). Simulated biomass was higher than measurement in 1982 at clay soil site. This is partially due to sampling occurring at full maturity, which was about 10–20 days after physiological maturity when the wheat reached its maximum biomass. Some senescent leaves, ripened spikes and broken tillers fall to the ground during this period, and are difficult to recover. The inaccurate simulation of grain yield (1993) and biomass (1982) may also reflect factors and/or events that are not considered in the model, such as heavy rainfalls, high winds, and hail damage, etc., which need to be improved with further study.

The simulation output of validated model was used to analyze the WU difference caused by soil texture. The simulated result showed that clay soil has higher WU than silt loam soil in most dry years, which agrees with research conducted in the same area as our study, by Lehane and Staple (1953)[37]. They also found that in a dry year, wheat grown on clay soil had a higher yield than those grown on silt loam soil. They attributed this effect to differences in crop access to the available moisture between soil textures. Wheat grown on coarser soil textures are able to readily extract stored soil water and spring precipitation, leading to lush early growth and heavy tillering.

Providing there is sufficient summer rains to replenish soil water, these crops have good growth and high yields. However, if soil water is not well replenished by summer rains, the wheat grown on coarser textured soils is unable to extract sufficient soil water to overcome water stress and so productivity suffers. In contrast, the wheat grown on fine textured soils are less able to extract soil moisture in the spring due to the water holding characteristics of the clay soil. Productivity is thus limited. However, if subsequent summer rains are insufficient to replenish soils water, the crops grown on clay soil are still able to access remaining soil water that they did not use in the spring. The results is that crops grown on clay soils are less sensitive to water use than those grown on coarser textured soils and produce higher yield in drought years[38]. Our simulation result concurs with these reports.

In conclusion, we found the CSM-CERES-Wheat model can be used as a tool to estimate the soil texture effects on crop productivity and WU. Crops grown on clay soil has better drought tolerance than those grown on silt loam in the climatic conditions of our study. We suggest that this work on estimating soil texture effects on WU and yield with simulation models be validated over a wider range of soil textures. The use of simulation models may lead to a better understanding of the crop–soil texture interactions, enabling crop breeders to analyze the performance of different crop genetics with regard to the broad soil textures that they target. For example, for cultivars targeted for production on clay soils, developing a variety that can extract more water from clay soil may improve both drought tolerance and production.

REFERENCES

1. Lobell, D. B. & Ortiz-Monasterio, J. I. Evaluating strategies for improved water use in spring wheat with CERES. *Agr. Water Manage.* 84, 249–258 (2006).
2. McConkey, B. G., Campbell, C. A., Zentner, R. P., Dyck, F. B. & lelles, F. Long-term tillage effects on spring wheat production on three soil textures in the Brown soil zone. *Can. J. Plant Sci.* 76, 747–756 (1996).
3. Zentner, R. P. *et al.* Cropping frequency, wheat classes and flexible rotations: Effects on production, nitrogen economy, and water

use in a Brown Chernozem. *Can. J. Plant Sci.* 83,667–680 (2003).

4. Campbell, C. A. *et al.* Long-term effects of cropping system and nitrogen and phosphorus fertilizer on production and nitrogen economy of grain crops in a Brown Chernozem. *Can. J. Plant Sci.* 85, 81–93 (2005).
5. Parton, W. J., Schimel, D., Cole, C. V. & Ojima, D. S. Analysis of factors controlling soil organic matter levels in Great Plains grasslands. *Soil Sci. Soc. Am. J.* 51, 1173–1179 (1987).
6. Carnol, M. *et al.* The effects of ammonium sulphate deposition and root sinks on soil solution chemistry in coniferous forest soils. *Biogeochemistry* 38, 255–280 (1997).
7. Chaudhari, S., Singh, R. & Kundu, D. Rapid textural analysis for saline and alkaline Soils with different physical and chemical properties. *Soil Sci. Soc. Am. J.* 72, 431–441 (2008).
8. Turner, N. C. Agronomic options for improving rainfall-use efficiency of crops in dryland farming systems. *J. Exp. Bot.* 55, 2413–2425 (2004).
9. Smith, W. N. *et al.* Assessing the effects of climate change on crop production and GHG emissions in Canada. *Agric. Ecosyst. Environ.* 179, 139–150 (2013).
10. Lobell, D. B., Ortiz-Monasterio, J. I., Addams, C. L. & Asner, G. P. Soil, climate, and management impacts on regional wheat productivity in Mexico from remote sensing. *Agr. Forest Meteorol.* 114, 31–43 (2002).
11. Liu, H. L. *et al.* Simulating water content, crop yield and nitrate-N loss under free and controlled tile drainage with subsurface irrigation using the DSSAT model. *Agr. Water Manage.* 98, 1105–1111 (2011).
12. Liu, H. L. *et al.* Using the DSSAT-CERES-Maize model to simulate crop yield and nitrogen cycling in fields under long-term continuous maize production. *Nutr. Cycl. Agroecos.* 89,313–328 (2011).
13. Liu, S., Yang, J. Y., Zhang, X. Y., Drury, C. F., Reynolds, W. D. & Hoogenboomd, G.Modelling crop yield, soil water content and soil temperature for a soybean rolled tile drainager conventional and conservation tillage systems in Northeast China. *Agr. Water Manage.* 123, 32–44 (2013).

14. Timsina, J. & Humphreys, E. Performance of CERES-Rice and CERES-Wheat models in rice-wheat systems: a review. *Agr. Syst.* 90, 5–31 (2006).

15. He, Y. *et al.* Spring wheat yield in the semiarid Canadian prairies: Effects of precipitation timing and soil texture over recent 30 years. *Field Crop. Res.* 149, 329–337 (2013).

16. He, Y., Wang, H., Qian, B., McConkey, B. & DePauw, R. M. How early can the seeding dates of spring wheat be under current and future climate in Saskatchewan, Canada? *PLoS ONE* 7,e45153 (2012).

17. Marchildon, G. P., Kulshreshtha, S., Wheaton, E. & Sauchyn, D. Drought and institutional adaptation in the Great Plains of Alberta and Saskatchewan, 1914–1939. *Nat. Hazards* 45,391–411 (2007).

18. Ayres, K. W., Acton, D. F. & Ellis, E. G. *The soils of the Swift Current Map Area 72J, Saskatchewan* (University of Saskatoon, Saskatoon, Saskatchewan, 1985).

19. Robertson, G. E. O. W. Wheat yields for 50 years at swift current Saskatchewan in relation to weather. *Can. J. Plant Sci.* 54, 625–650 (1974).

20. McConkey, B. G., Curtin, D., Campbell, C. A., Brandt, S. A. & Selles, F. Crop and soil nitrogen status of tilled and no-tillage systems in semi-arid regions of Saskatchewan. *Can. J. Plant Sci.*82, 489–498 (2002).

21. Jame, Y. W. & Cutforth, H. W. Simulating the effects of temperature and seeding depth on germination and emergence of spring wheat. *Agr. Forest Meteorol.* 124, 207–218 (2004).

22. Ritchie, J. T. [Wheat phasic development.]. *Modeling Plant and Soil Systems, Agronomy Monograph* [Hanks, R. J. (ed.), Ritchie, J. T. (ed.)] [31–54] (ASA-CSSA-SSSA, Madison, Wisconsin, 1991).

23. Wang, H. *et al.* Predicting the time to 50% seedling emergence in wheat using a Beta model.*NJAS-Wagen. J. Life. Sc.* 57, 65–71 (2009).

24. Thornton, P. E., Hasenauer, H. & White, M. A. Simultaneous estimation of daily solar radiation and humidity from observed temperature and precipitation: an application over complex terrain in Austria. *Agr. Forest Meteorol.* 104, 255–271 (2000).

25. Saxton, K. E. & Rawls, W. J. Soil water characteristic estimates by texture and organic matter for hydrologic solutions. *Soil Sci. Soc. Am. J.* 70, 1569–1578 (2006).
26. Gijsman, A. J., Thornton, P. K. & Hoogenboom, G. Using the WISE database to parameterize soil inputs for crop simulation models. *Comput. Electron. Agr.* 56, 85–100 (2007).
27. Boote, K. J. [Decision Support System for Agrotechnology Transfer Version 3.0.]. *Concepts for calibrating crop growth models* [Hoogenboom, G. (ed.), Wilkens, P. W. (ed.), Tsuji, G. Y. (ed.)] [179–200] (University of Hawaii, Honolulu, 1999).
28. Wang, H., McCaig, T. N., DePauw, R. M., Clarke, F. R. & Clarke, J. M. Physiological characteristics of recent Canada Western Red Spring wheat cultivars: Yield components and dry matter production. *Can. J. Plant Sci.* 82, 299–306 (2002).
29. Willmott, C. J. On the validation of models. *Phys. Geogr.* 2, 184–194 (1981).
30. Meng, L. & Quiring, S. M. A comparison of soil moisture models using soil climate analysis network (SCAN) observations. *J. Hydrometeorol.* 9, 641–659 (2008).
31. Hoogenboom, G. *et al.* [Decision Support System for Agrotechnology Transfer Version 4.0.].*Overview* [Hoogenboom, G. (ed.) *et al.*] (University of Hawaii, Honolulu, 2003).
32. Kersebaum, K. C. *et al.* Long-term simulation of soil-crop interactions in semiarid southwestern Saskatchewan, Canada. *Eur. J. of Agron.* 29, 1–12 (2008).
33. Saseendran, S. A., Nielsen, D. C., Ma, L. & Ahuja, L. R. Adapting CROPGRO for simulating spring canola growth with both RZWQM2 and DSSAT4.0. *Agron. J.* 102, 1606–1628 (2010).
34. Kröbel, R. *et al.* Nitrogen and phosphorus effects on water use efficiency of spring wheat grown in a semi-arid region of the Canadian prairies. *Can. J. Plant Sci.* 92, 573–587 (2012).
35. Rötter, R. P., Carter, T. R., Olesen, J. E. & Porter, J. R. Crop-climate models need an overhaul. *Nat. Clim. Change* 1, 175–177 (2011).
36. Webber, H., Gaiser, T. & Ewert, F. What role can crop models play in supporting climate change adaptation decisions to enhance food security in Sub-Saharan Africa? *Agr. Syst.*127, 161–177 (2014).

37. Lehane, J. J. & Staple, W. J. Water retention and availability in soils related to drought resistance. *Can. J. Plant Sci.* 33, 265–273 (1953).
38. Lehane, J. J. & Staple, W. J. Effects of soil moisture tensions on growth of wheat. *Can. J. Plant Sci.* 42, 180–188 (1962).

Chapter 7

Soil Fungal Resources in Annual Cropping Systems and Their Potential for Management

Walid Ellouze[1], Ahmad Esmaeili Taheri[1,2], Luke D. Bainard[1], Chao Yang[1,2], Navid Bazghaleh[1,3], Adriana Navarro-Borrell[1,3], Keith Hanson[1], and Chantal Hamel[1]

[1]Semiarid Prairie Agricultural Research Centre, Agriculture and Agri-Food Canada, P.O. Box 1030, 1 Airport Road, Swift Current, SK, Canada S9H 3X2

[2]Department Food and Bioproduct Sciences, College of Agriculture and Bioresources, University of Saskatchewan, 51 Campus Drive, Saskatoon, SK, Canada S7N 5A8

[3]Department of Soil Science, College of Agriculture and Bioresources, University of Saskatchewan, 51 Campus Drive, Saskatoon, SK, Canada S7N 5A8

ABSTRACT

Soil fungi are a critical component of agroecosystems and provide ecological services that impact the production of food and bioproducts. Effective management of fungal resources is essential to optimize the productivity and sustainability of agricultural ecosystems. In this review, we (i) highlight the functional groups of fungi that play key roles in agricultural ecosystems, (ii) examine the influence of agronomic practices on these fungi, and (iii) propose ways to improve the management and contribution of soil fungi to annual cropping systems. Many of these key soil fungal organisms (i.e., arbuscular mycorrhizal fungi and fungal root endophytes) interact directly with plants and are determinants of the efficiency of agroecosystems. In turn, plants largely control rhizosphere fungi through the production of carbon and energy rich compounds and of bioactive phytochemicals, making them a powerful tool for the management of soil fungal diversity in agriculture. The use of crop rotations and selection of optimal plant genotypes can be used to improve soil biodiversity and promote beneficial soil fungi. In addition, other agronomic practices (e.g., no-till, microbial inoculants, and biochemical amendments) can be used to enhance the effect of beneficial fungi and increase the health and productivity of cultivated soils.

INTRODUCTION

Microorganisms are involved in fundamental processes such as soil formation and nutrient cycling and can be seen as the cornerstone of the biosphere. They are an essential link between soil nutrient availability and plant productivity as they are directly involved in the cycling of nutrients through the transformation of organic and inorganic forms of nutrients. Certain microorganisms, in particular those interacting physically with plants in the rhizosphere, can also influence plant productivity negatively by causing disease or positively by enhancing plant growth.

In a world of seven billion people, the production of food and biofuel occupies an important proportion of the Earth's surface and therefore cropping systems must be efficient and sustainable. In light of the importance of soil microorganisms in the productivity of agroecosystems, the management of beneficial soil microbial diversity emerges as a new strategy for crop production in a changing world. This review considers the factors affecting the fungal resources relevant to agriculture and explores avenues toward the management of these resources to improve the efficiency of crop production. We propose a model where the plant is the key to the management of soil fungal resources and where the fungi living in close association with plant roots constitute the manageable resource (Figure 1).

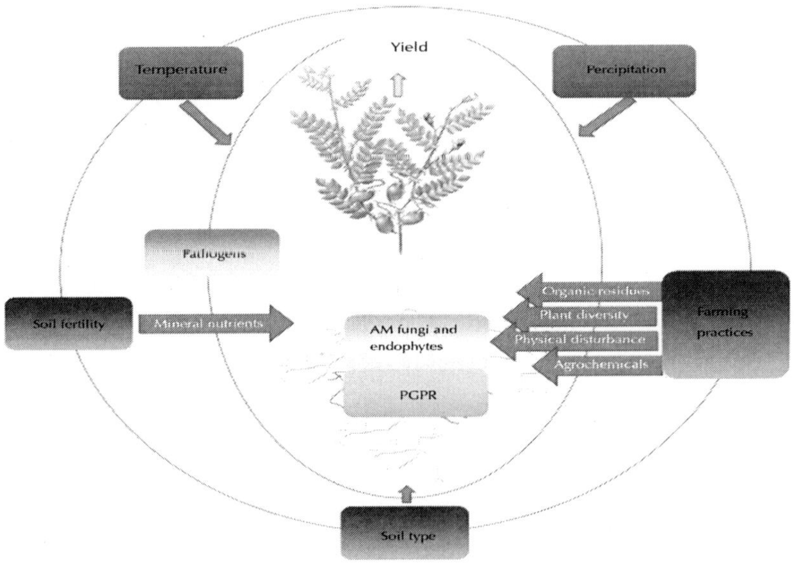

Figure 1: Graphical overview of the relationships between plant-associated microbial diversity, crop yield, and environmental conditions in agroecosystems as influenced by management.

In our view, arbuscular mycorrhizal (AM) fungi and fungal endophytes are the fungi that should be the target of management. We will review these key soil fungal groups, the plant mechanisms regulating them, and present different ways that could be used to improve soil health and, consequently, the efficiency of annual cropping systems. Although the concepts presented in this review are often relevant to all crops and production systems, they will be primarily illustrated with reference to dry land crops and cropping practices used in the cool and subtropical climates.

IMPORTANT SOIL FUNGI IN AGRO-ECOSYSTEMS

Arbuscular Mycorrhizal Fungi

AM fungi are ubiquitous in terrestrial ecosystems and form a symbiotic relationship with the roots of most plants [1]. They are obligate biotrophs requiring a plant partner for their carbon supply and are unable to complete their reproductive cycle without a host plant [2]. Initiation of the symbiosis can occur through the colonization of plant roots by germinating spores, hyphae, or infected root fragments [3]. Upon colonization, AM fungi form different functional structures in the root cortex of the host plant including arbuscules and hyphal coils (primary sites of nutrient exchange), vesicles (storage structures), and spores (reproduction) [1]. Through the AM symbiosis, the host plant is connected to extensive hyphal networks in the soil [4].

The primary function of the AM symbiosis involves a bidirectional transfer of carbon from the plant in exchange for soil-derived nutrients from the fungal partner [1]. Extensive networks of extraradical mycelium in the soil enable the fungus to uptake and rapidly translocate nutrients to intraradical arbuscules and hyphal coils and into the plant, thereby increasing the availability of soil nutrients in the soil to the host plant [1]. In addition, AM fungi can provide other functional benefits to the host plant such as improved water relations [5] and protection from pathogens and herbivores [6, 7]. The AM association is usually mutualistic, but evidence does suggest that it can range from parasitic to mutualistic [8].

AM fungi are also involved in several important ecosystem processes. They have a direct effect on plant productivity and have been shown to influence plant diversity and community structure [9–11]. In addition, the extensive mycelial networks produced by AM fungi coupled with the secretion of glomalin have a beneficial impact on soil health by improving the structural stability, quality, and water retention of soil [12,13]. AM fungi also play an important role in the cycling of major elements such as carbon (C), phosphorus (P), and nitrogen (N) [14]. From an agroecological perspective, the functions and ecological services provided by AM fungi reveal the important impact these symbiotic organisms have on the productivity and sustainability of agricultural systems [15–17].

There are various abiotic and biotic factors that influence the distribution, growth, and function of AM fungi. These include abiotic factors such as soil chemistry (e.g., pH, nutrient availability, and pesticides [18, 19]), climatic variables (e.g., temperature, light, and precipitation [20–22]), and soil structure and stability [23, 24]. Biotic factors are primarily linked to the composition of the plant community as several studies have found that the diversity and assembly of AM fungal communities are strongly influenced by the plant community [22, 25–27]. Other biotic factors that have been shown to influence AM fungi are root predators [28], plant parasites [29], and herbivores [30]. Many of these abiotic and biotic factors are interrelated and interact synergistically to influence the habitat and in turn the composition and functioning of AM fungal communities.

In agricultural systems, many of these abiotic and biotic factors are modified by management techniques, which strongly impact AM fungal communities. Studies have shown that practices such as tillage and fallow [31, 32], monoculture cropping [33], and fertilization [34] all negatively influence the abundance and diversity of AM fungi. In general, agroecosystems have a lower AM fungal diversity compared to natural ecosystems [35] and this loss of diversity appears to be correlated with management intensity [36, 37].

Fungal Endophytes

Two important groups of non-AM fungi associated with plant roots are functionally defined as pathogens and endophytes. Both fungal endophytes and pathogens can colonize plant tissue, but, in contrast to endophytes, pathogens are able to cause disease in plants [38]. Pathogenicity is not exclusive to fungi, but in agricultural systems most plant diseases are caused by fungal pathogens [39]. Fungal pathogens have attracted much research attention because they are responsible for very important yield losses. Fungal pathogens are unwanted in agroecosystems and agronomic practices are aimed at controlling their abundance and their impacts.

Endophytic fungi are a group composed of very heterogeneous fungi that have been divided into two major groups: clavicipitaceous and nonclavicipitaceous endophytes [40]. Clavicipitaceous endophytes are a small group of fungi usually transmitted through seeds and that colonize the shoots of some grass species [40, 41]. The nonclavicipitaceous endophytes are a very diverse group of fungi (primarily ascomycetous) sharing the capacity to colonize the root systems of a wide range of plant lineages and which often have dark and septate hyphae [40]. While little is known about the ecology and functionality of endophytic fungi, a growing number of reports have revealed the beneficial services provided by endophytic fungi to host plants. The potential for commercial application of mutualistic endophytes with biocontrol abilities has promoted research in this field and several bioproducts for the control of plant diseases are already commercially available [42].

Many endophytic fungi have been reported to protect plants against diseases. For example, inoculation withBeauveria bassiana protected cotton and tomato against the pathogens Rhizoctonia solani and Pythium myriotylum [43]. Trichoderma atroviride and Epicoccum nigrum also protected potato against Rhizoctonia solani [44]. Trichoderma is a genus well known for having biocontrol activity against pathogenic species and some Trichoderma isolates are formulated and used as inoculants for the control of several plant diseases like onion white rot, Fusarium wilt of chickpea, and Fusarium crown and root rot of tomato [42, 44–48]. Different mechanisms are suggested to explain the protection of plants by their fungal endophytes [49] including competition for niche occupation and resource utilization [43], direct interaction [50, 51], or induced systemic resistance [43, 52].

Some fungal endophytes can also protect plants against abiotic stress created by drought [53], salinity [54], or toxic levels of metal [55], while others were reported to promote plant growth [52–54, 56]. The production of plant hormones and growth regulators appears to be an important mechanism by which fungal endophytes improve plant growth and yield under stressful conditions [54].

Accumulating evidence indicates a nutritional effect of soil fungal endophytes on their host plant (e.g., [57,58]). Solubilisation of soil phosphorus appears to be involved in the improved plant P uptake mediated by fungal endophytes [54, 59]. In addition, enhanced mineralization is suggested to explain the role of fungal endophytes in plant nitrogen nutrition [49].

Understanding population dynamics and community structure of fungi in agricultural systems is necessary to minimize the damage from pathogens and optimize the benefits of mutualistic fungi. In addition to natural environmental fluctuations, anthropogenic activities can drastically affect fungal communities. Potent pathogens are carried across continents [60] and climate warming will shift the host range and fruiting date of some important fungi [61, 62]. In agroecosystems, cropping practices have profound and immediate impacts on the soil fungal community by modifying environmental factors such as soil pH, fertility, moisture, and plant cover. Among soil properties, pH is known as a major factor shaping the community of root-associated fungi [63, 64]. Soil nutrient availability and organic matter content are also thought to influence root endophyte diversity [65–67]. However, host preference is the most important factor in plant-fungal relationships [52, 63, 68] and crop selection likely has the strongest effect on fungal endophyte community composition in agroecosystems.

MECHANISMS OF PLANT CONTROL OVER FUNGI

Plants coexist with a wide variety of beneficial and pathogenic fungi at all stages of their life. Plants actively interact with fungi using numerous mechanical and biochemical tools [131] and have evolved sophisticated strategies to shape the structure and function of their fungal environment [132]. Rhizodeposition is the process through which plant roots release organic and inorganic compounds that modify the physical, chemical, and biological properties of their soil environment [133–135]. Plant roots release a wide array of compounds that act as nutrient sources for soil fungi and as highly specific chemicals involved in diverse biological interactions [136, 137]. The secretion of carbon compounds derived from cortical and epidermal cells stimulates the proliferation of fungi outside, on the surface, and inside the roots [134]. An abundance of fungal growth on the root creates a barrier inhibiting the relative growth of pathogenic microorganisms through interspecific competition.

Several chemical pathways involved in the communication between plants and soil fungi have been identified and are illustrated in Figure 2. Phenolic compounds play key roles in presymbiotic stages of the AM symbiosis. They stimulate AM hyphal growth and branching [138]. Root symbioses are tightly controlled interactions. The extent to which root tissues are colonized by AM fungi and rhizobia is subjected to autoregulatory mechanisms preventing excessive colonization of the roots by the microsymbionts, thus preserving the symbiotic nature of the associations [139].

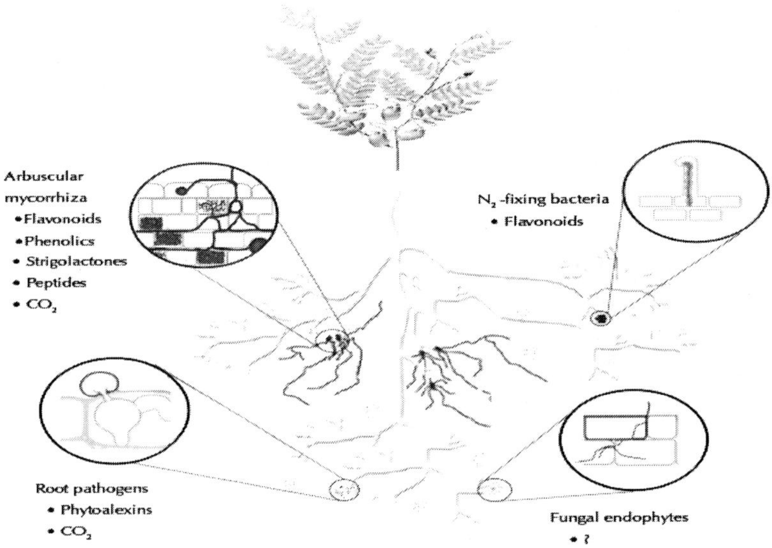

Figure 2: General overview of the bioactive phytochemicals involved in interactions between plants and soil microorganisms.

Plant hormones play a major role in the complex signalling and regulatory processes controlling plant-fungus interactions [140]. These include salicylic acid, ethylene, jasmonic acid, abscisic acid, gibberellic acid, auxin, cytokinin, strigolactones, and brassinosteroids. Salicylic acid is associated with the control of biotrophic plant pathogens while ethylene and jasmonates are involved in plant defence against necrotrophs [140]. Strigolactones are exuded into the rhizosphere under harsh environmental conditions and are known to stimulate hyphal branching of AM fungi and generally inhibit the growth of pathogenic fungi [141–143].

Plant proteins are also involved in interactions with soil fungi. Tryptophan dimers secreted from Bahia grass roots acted as a signal, stimulating the growth of AM fungal hyphae, under water-limiting conditions [144]. Peptides with hormonal activity are a component of the defence mechanism of plants [140]. Plant roots also secrete a wide spectrum of antimicrobial proteins such as chitinases that disrupt the cell wall and suppress the growth and function of pathogenic fungi [145, 146]. Extensin and other proteins identified in a root extract appeared to be involved in the suppression of AM fungal spore germination [147]. Furthermore, several types of volatile organic compounds (VOC) were found to trigger responses in insects but also to suppress the growth of pathogenic fungi, in particular Fusarium spp. [148, 149]. Plant-fungus interactions are highly complex and involve hormonal, mechanical, and biochemical factors.

Plants are more than a mere source of nutrients for soil fungi. They have coevolved with specific fungi and specific soil fungal communities, which led to the emergence of various lifestyles and forms of coexistence in the plant kingdom. For example, plants from the Fabaceae, such as pea, bean, and lentil, are associated with AM fungi [150]. Wheat, barley, rye, and oat are members of the Poaceae and they associate with AM fungi [151, 152], but as members of the subfamily Pooideae, they rarely respond to the symbiosis [153]. The Brassicaceae, including oil seed canola or mustard, do not associate with AM fungi or rhizobia [154].

Plants influence soil fungal diversity. The cultivation of mycorrhizal crops increases the inoculum density, which promotes the formation of mycorrhizal symbioses in the following seasons. Research has revealed that when a mycorrhizal crop is cultivated in rotation after a nonmycorrhizal crop, root colonization and symbiotic contributions to plant growth are delayed as a result of decreased levels of inoculum in the soil [111]. The genotypes and species of these broad taxonomic groups of plants have different phytochemistry [147, 149] and influence the soil microbial communities in slightly different ways [155].

MANAGEMENT OF SOIL FUNGAL RESOURCES

Management through Genetic Selection of Plants

Technologies for agriculture have emerged from research on the biochemistry of plant-microbe regulation. The use of formulations of flavonoids or lipochitooligosaccharides at seeding now enhances crop production in fields of the Canadian prairies and elsewhere through the use of products such as PulseSignal II or Optimize (Novozymes BioAg Group). The mechanisms plants implement to manage their microbial environment are complex [131, 132] and as difficult to manipulate as they are finely regulated. The intraspecific variation observed in the profile of plant signaling phytochemicals [147, 149] and concurrent fungal environment [155] suggests the possibility of selecting crop plants with special compatibility with beneficial fungi. The selection of plant genotypes resistant to pathogens has already led to important progress in phytoprotection [156] and points toward plant management as a key to managing soil fungal resources in agroecosystems. Selection of plant genotypes that have favourable compatibility with beneficial soil fungi is possible, as shown by variation in the compatibility of certain genotypes with beneficial fungi that were found in the studies listed in Table 1.

Table 1: Reports of intraspecific genetic variation in the ability of crop plants to host beneficial fungal endophytes, a necessary condition for genotype selection in genetic improvement programs

Microorganism	Type and function	Host plant	References
AM fungi	Symbiotic soil fungi improving the ability of host plants to extract soil nutrients	Wheat (Triticum spp.)	[69–82]
		Barley (Hordeum vulgare L.)	[83]
		Triticale (×Triticosecale)	[82]
		Oats (Avena spp.)	[84]
		Maize (Zea mays L.)	[85–90]
		Rice (Oryza sativa L.)	[91, 92]
		Soybean (Glycine max (L.) Merr.)	[88, 93]
		Onion (Allium spp.)	[94, 95]
		Tomato (Lycopersicon esculentum Mill.)	[96]
		Peanut (Arachis hypogaea L.)	[97]
		Marigold (Tagetes spp.)	[98]
		Pepper (Capsicum annuum L.)	[99]
Acremonium	Fungal shoot endophyte increasing plant vigor, resistance to insects, and modifying water relations	Wheat (Triticum spp.)	[100]
Neotyphodium	Fungal shoot endophyte improving plant tolerance to stress	Wheat (Triticum spp.)	[100]

Growing crop varieties with improved compatibility with beneficial soil fungi can be a powerful way to manage soil fungi and a good strategy to enhance soil nutrient use efficiency in agroecosystems. Some studies suggest that modern breeding programs conducted in highly fertilized systems may have produced cultivars with a high level of dependence on fertilizer and a diminished capacity to form symbiotic relationships with beneficial soil fungi [69–71, 100, 157]. However, this hypothesis was disproved by a meta-analysis evaluating the importance of the year of release on mycorrhizal responsiveness, AM fungal root colonization, and P efficiency [158]. There is little evidence to support a negative impact of plant breeding on AM formation and function. In fact, the prolific growth of AM fungi that can be seen in the rhizosphere of certain recent cultivars [72] could suggest that modern plant breeding approaches have improved the microbial associations with crop roots.

Plant genotypes differentially influence the soil microbial communities of agricultural fields [155]. Mixtures of cultivars have led to yield stability over a range of environmental conditions and sustained higher productivity than monocultures [159]. These effects were attributed to crops maintaining health-promoting soil microbial communities [109]. Mixtures of cultivars create diversified niches that maintain a higher diversity of beneficial soil microorganisms with host preference [160] and functional complementarity [124].

Overall, breeding crop varieties with an improved ability to interact with beneficial soil fungi appear to be a logical approach to enhance crop yield. Targeting plant genes responsible for beneficial interactions with soil fungi should improve the nutrient efficiency of crops and reduce the environmental impacts of fertilization, as well as farm input costs, leading to more sustainable production systems.

Management through Rotation

Certain agronomic practices are designed to manage biodiversity in the agroecosystem by enhancing diversity and repressing pests and disease outbreaks (Table 2). Among these practices, rotating crops is one of the more traditional and effective ways to diversify the microbial community, reduce the impact of diseases and weeds [101], and thus increase yields. The value of a cropping system depends on a number of factors including the genotype and crops included in the rotation [102], the sequence and frequency of the crops [103], the length of the rotation [161], the management history [162], and soil characteristics [163]. Overall, these factors impact the soil microbial community in different ways.

Table 2: General effects of agronomic practices on soil fungal diversity and abundance, disease incidence, soil fertility, crop nutrient use efficiency, and crop growth and yield

Source of effects	Biodiversity level	**Crop growth and productivity**	**Disease, pests and pathogens**	**Microbial abundance**	**Soil fertility**	**Nutrient use efficiency**	**References**
Biodiversity management							
Crop rotation	+[a]	+	−	+			[101–108]
Cultivar mix	+	+	−	+			[103, 104, 109]
Intercropping	+	±	−	+			[106]
Cover cropping	+	±	−	+			[106, 107, 110]
Nonmycorrhizal crops	−		+			−	[111]
Transgenic crops	0	±	−	0			[112–117]
Pesticide use	0	+	−	0		−	[118–121]
Weed control	−	+		−		+	[118, 120]
Inoculants		±	+		+	+	[80, 91, 97, 98,122, 123]
Soil management							

Organic amendments	+	+		+	+	±	[102, 124, 125]
Nitrogen fertilizers	±	+		+	+	−	[126, 127]
Mineral fertilization		+		±	+	−	[120, 126]
Tillage	±	±	±	±	±	±	[31, 128–130]

a+ (positive to no effects), 0 (negligible effects), − (negative to no effects), and ± (variable effect).

Intercropping systems and crop rotations offer opportunities for a better management of soil fungi. Using mixtures of different cereal genotypes [104, 109] or crops such as wheat, barley, canola [105], clover, and alfalfa [106] can enhance productivity by reducing weeds and disease incidence at the system level. Also, changes in the frequencies of cultivars [103, 104] over time can influence the incidence of stem and root rot diseases in the rotation system and enhance yield stability. For example, corn grain yield can increase linearly in relation to the number of crops included in the rotation up to twice the yield of the monocrop when three rotation crops and three cover crops are included in the cropping system [107]. Certain crops in the rotation are better than others and it can be complicated to determine what the optimal rotation sequence to maximize benefits is [103]. Soil factors are also important to consider in the design of rotation sequences (e.g., soil-water stable aggregation, soil organic C, and the carbohydrate composition of the surface layer) as these parameters also affect the abundance, diversity, and distribution of the fungal community [108]. In most cases, monoculture negatively affects microbial biomass and diversity [164, 165]. Diversifying the crops used in rotation increases the taxonomic and functional diversity of soil fungal communities [166]. In addition, microbial activity and substrate utilization are significantly affected by crop rotation [110]. Different crops provide different organic residues, which can result in a diverse food base that promotes fungal diversity and activity and increases soil fungal biomass and N mineralization [167]. Interestingly, the biochemical composition of some plant tissues can modulate the fungal associations. Plants of the Poaceae are particularly rich in pentoses, which are the main energy source of soil fungi. So it is not surprising that many fungi are associated with cereals.

Diversifying crop rotations also decreases disease pressure in agroecosystems by disrupting the life cycle of pathogens associated with a particular crop or plant genotype. The length and level of crop diversity are key factors for the success of a cropping system. Short rotations are more susceptible to diseases and produce lower yield than longer rotations [161]. Other factors to consider in the design of crop rotation systems include the ability of plant pathogens to use alternative host plants or remain dormant in the soil for long periods [168] and allelopathy and autotoxicity of crops [161]. Selecting plants that are not alternate hosts for pathogenic fungi in other components of the rotation is important to reduce yield losses due to diseases. However, some pathogens can persist in the soil for several years as spores or other dormant structures, in absence of a host plant [168]. In addition, monocultures negatively affect fungal biodiversity by selecting for virulent pathogens, which then have a competitive edge and increase disease severity. In a continuous-pea rotation grown in the Canadian prairie, severe Fusarium root rot injury was related to a reduced soil microbial community and lower abundance of beneficial Gram positive bacteria and AM fungi [165]. In some cases, continuous cropping has increased the abundance of antagonistic microorganisms and reduced pathogen populations, mitigating the impact of take-all in wheat [102], but as a general rule, at least three and possibly more crops should be included in cropping systems [161]. The inclusion of cover crops in cropping systems is particularly effective in reducing disease incidence [110].

In semiarid cold and subtropical steppes, farmers have traditionally grown cereals in alternation with summer fallow. This consists of keeping the soil bare using tillage or herbicides during a growing season. In the last two decades, broadleaf crops such as field pea, lentil, chickpea, canola, and mustard were introduced in wheat-based rotations in the semiarid area of the Canadian prairie to replace summer fallow, which lost relevance with the development of no-till systems for soil moisture conservation [107, 169]. Crop diversification with broadleaf crops, especially pulses, has the benefit of increasing grain yield and protein content of the wheat crops following in rotation, partially due to residual soil N from biological fixation [103].

Canola and mustard are nonmycorrhizal plants that do not associate with rhizobacteria. These crops also require the use of more N and S fertilizers; however, the productivity and value of these crops compensate for the larger investment in fertilizers. Despite the economic benefit of these crops, having nonmycorrhizal plants in the crop rotation may reduce AM fungal populations and delay mycorrhizal formation in the following crop [170, 171], which may impact AM dependent crop plants. Clearly, there are many factors to consider in the design of ecologically sustainable and economically viable crop rotation systems.

Management through Biochemical Amendments

The use of biologically active chemicals is an alternative approach to managing the structure and function of soil fungal communities. Plants naturally release a wide spectrum of bioactive phytochemicals that modify their microbial environment. The phytochemicals contained in plants vary with the species, genotype, tissue, physiological stage, and environmental conditions [147, 149, 172, 173]. The application of plant tissues containing certain phytochemicals as dried organic amendment or green manure can effectively reduce the inoculum of soil borne plant pathogens and stimulate the growth of beneficial fungi. For example, incorporating the tissues of certain legumes into infected soils has shown the potential to control parasitic nematodes and reduce gall number in tomato [174]. These legumes contain bioactive phytochemicals that negatively impact plant-parasitic nematodes. Plants of the Brassicaceae contain glucosinolates and have long been known for their activity against fungal pathogens. Brassica napus seed meal applied to orchard soil reduced the infection by fungal pathogens (Rhizoctonia spp.) and parasitic nematodes (Pratylenchus spp.) of apple roots [175]. The control of Rhizoctonia root rot of apple by B. napus was attributed to the modification of the bacterial community structure and the induction of plant systemic resistance [175]. This suggests that stimulating soil fungal communities by the addition of bioactive amendments may be an effective way to manage soil fungal communities and control pathogens [176].

The production of bioactive VOC by plants can trigger responses in the organisms surrounding them and inhibit certain pathogens. Changes in the profile of VOC by plants are generally a response to pathogenic invasion. For example, the profile of VOC from chickpea was correlated with Ascochyta blight severity [149]. The VOC of chickpea, in particular trans-2-hexenal and 1-hexanol, were much more potent against the causing agents of Fusarium head blight than wheat VOC [149]. This provided an explanation for the susceptibility of wheat and the resistance of chickpea to these pathogens [149]. Selection of genotypes based on VOC production may be a strategy to increase disease resistance in crop rotations.

Management through Inoculation and Soil Management

Rhizobial inoculants have been used in agricultural systems for decades and are proven efficient tools to manage beneficial soil microbial diversity. Inoculation of crops with selected plant growth promoting microbial strains (e.g., PGP rhizobacteria and AM fungi) is a strategy that can easily be integrated into cropping systems [177, 178]. Although simple, inoculation of crops can be unreliable. Competition among microorganisms in the soil system can be intense and introduced organisms may not live long enough to produce the desired effect, especially if their niche is not unique. The combination of inoculation along with certain agronomic practices may increase the probability of beneficial effects from inoculants. Practices that modify the soil environment in a way that benefits the introduced microorganisms may increase the value of inoculants.

Soil properties can modify the influence of fungi on plants and management practices that modify soil properties could be used to maximize the beneficial effects of inoculants. Because soil organic matter (SOM) controls many soil properties [179], the management of SOM appears to be a key to managing soil microorganisms. Amending the soil with organic materials and adopting conservation tillage practices are strategies that most effectively influence SOM.

Fresh organic matter and manure have a stronger effect on microbially mediated soil structuration than stable organic matter, but the effect of the latter is long-lasting particularly if large amounts are applied. Organic amendments contain energy and nutrients favouring fungal proliferation and are also rich in functional groups that can adsorb nutrients and retain water. This increases the soil nutrient pool and soil moisture levels, which further supports the growth and function of plants and fungi.

The addition of fresh organic material can immediately boost the performance of inoculants in annual cropping systems. For example, the addition of manure to soil improved the contribution of fungal biocontrol agents to plant health [122] and the plant-growth-stimulating effects of AM fungi [123]. Organic amendments benefit the microorganisms using them by providing a source of nutrients and energy, but the positive effect of organic amendments has also been attributed to their impact on soil physical quality [122]. The stimulation of fungal growth by organic amendments triggers the production of aggregate-stabilizing fungal filaments and exopolysaccharides that structure the soil matrix, which increases its porosity and has a positive effect on gas exchange and water infiltration and retention.

In regions where sources of organic amendments are not readily available such as the intensive grain producing steppes, no-tillage practices are effective methods for soil moisture conservation and increasing SOM levels [180]. Soil tillage has tremendous effects on soil physical and biological properties by homogenizing the soil matrix and stimulating mineralization, which in the long term reduces the level of SOM [180]. Consequently, the influence of no-till practices on soil physical properties is in many ways similar to the influence of organic amendments. Soil aggregates are conserved in the absence of tillage and the soil is well structured and porous. The organic matter is preserved in stable aggregates favouring SOM accumulation, which further improves soil porosity, aeration, and water infiltration and retention. The heterogeneity of the soil in the absence of tillage leads to the development of a variety of niches allowing the establishment of highly diverse microbial communities. However, the effect of no-till on SOM accretion is slow and develops through decades after the abandonment of intensive tillage practices [180]. The addition of organic amendments to soils with suboptimal physical properties is useful in accelerating the establishment of soil physical conditions hospitable to beneficial microorganisms.

While the combined use of organic amendments and inoculants can increase the performance of PGP microorganisms in cultivated fields, excessive rates of organic amendments may also be inhibitory to certain PGP microorganisms. For example, high concentrations of compost can inhibit AM fungi, whereas low rates are beneficial [124]. In addition, amendments used to create conditions favourable to crop plants may negatively impact the beneficial microbial associates of plants that are adapted to soil conditions suboptimal for production. This was shown to be the case for certain AM fungi, which had a reduced ability to colonize their host after a saline-alkali soil was amended with gypsum [181]. Although the soil conditions conducive to biological activity and biodiversity may be suboptimal for certain microorganisms with PGP activity, the maintenance of soil physical quality should favour the survival and functional activities of most PGP microorganisms introduced in agroecosystems through inoculation.

INFLUENCE OF AGROCHEMICALS ON SOIL MICROORGANISMS

Managing the soil environment through the use of agrochemicals is often secondary to the primary goal of these products, but they are widely used and can strongly influence soil microbial communities. On the Canadian prairie, 73% of the land in crop production receives chemical inputs in the form of pesticides and/or inorganic fertilizers [182]. Most production requires fertilizer with inputs of 1.3 million metric tonnes of N and 0.48 million metric tonnes of P applied annually [183]. With this level of inputs going onto the soil it is important to understand and manage the effects these chemicals have on the soil environment.

There are many different fertilizer formulations available and some include amendments that directly affect and inhibit microbial activity [126]. Soil pH can be affected by different factors including the use of inorganic N fertilizers. The majority of the N applied is in the form of granular urea or anhydrous ammonia, both of which have been found to be less acidifying to soil than ammonium sulphate or ammonium phosphate formulations. The level of acidification resulting from ammonium fertilizer varies with soil characteristics and cropping systems [184], but there is considerable evidence of soil acidification due to N fertilizer use in the Canadian prairies despite the high buffering capacity of these soils [127, 185, 186]. The drop in pH can be alleviated by liming the soil, but the associated costs limit the use of this practice [187]. In general, long term N use lowers soil pH and in turn has a negative impact on certain soil microbial groups, especially actinomycetes and denitrifying bacteria. In general, fungi can tolerate a wider range of pH than bacteria [188]. Lower pH does not appear detrimental to fungi and may sometimes increase their abundance [127].

With the exception of soil fumigants and certain fungicides, pesticides appear to have a limited effect on soil fungi [189]. Recent studies have demonstrated that pesticides have a minimal effect on soil fungi when they are applied at the recommended doses [190]. However, pesticides may influence the function and ecological processes associated with the soil fungal community. For example, there is some evidence that pesticides can effect soil biochemical reactions, especially related to nutrient cycling [118, 191]. In addition, the application of fungicides against foliar disease influences not only the production of VOC in the aboveground tissues, but also the production of these antimicrobial phytochemicals in the roots [149]. As a result, foliar applied fungicides can significantly affect plant-pathogen interactions in the rhizosphere. The widely used herbicide glyphosate can also modify the structure of rhizosphere fungi under certain cropping practices [119].

Since fertilizers and pesticides are commonly used together in conventional cropping systems, it is important to understand the interactive effects of these agrochemicals. A study in the Canadian prairies found that, in the short term, fertilizers and herbicides have beneficial or minimal effects on soil microbiological characteristics [120]. However, over time some deleterious effects on soil microorganisms and their associated biological processes were observed indicating the cumulative effect of repeated applications of agrochemicals [120]. Meanwhile, other studies have reported interactive effects of pesticides and soil fertility on soil microbial communities. For example, herbicides had a more pronounced effect on soil microbial community structure in soils with low fertility [192] and in crops not fertilized with N [119]. Furthermore, fertilization can influence the degradation of pesticides and modify their nontarget effects on soil microbial communities [121].

The influence of agrochemicals on important soil fungi is complex and difficult to predict, further increasing the difficulty involved in the management of soil fungal resources. Agrochemicals are abundantly used in annual crop production systems and are considered a necessity to achieve desired crop yields. Future research should focus on optimizing pesticide and fertilizer applications that promote beneficial soil fungi and their associated biological processes to encourage more sustainable agroecosystems that are less dependent on conventional agrochemicals.

CONCLUSIONS

The soil fungi that have the strongest influence on plants reside in the rhizosphere and it appears that plants can be used to manipulate these fungi in order to improve soil health and the efficiency of annual cropping systems. In this context, the traditional practice of crop rotation can be used as a basic strategy to increase diversity in the rhizosphere and prevent the build-up of pathogens. Future approaches to complement crop rotations will likely include the use of cultivars with specific compatibilities with beneficial fungi. In addition, biotechnologies based on the use of bioactive phytochemicals and fungal inoculants are currently available and are being diversified and refined. Combining inoculation with practices that create conditions favourable to the survival and activity of the desirable fungi will be an effective strategy to increase the value of inoculants. Despite the complexity of the soil ecosystem, it is possible to manage soil fungal diversity in order to promote more sustainable and productive agroecosystems. As global change dictates the need for more efficient cropping systems, the management of beneficial fungi offers many opportunities.

AUTHORS' CONTRIBUTION

Walid Ellouze, Ahmad Esmaeili Taheri, Luke D. Bainard, Chao Yang, Navid Bazghaleh, Adriana Navarro-Borrell, and Keith Hanson contributed equally to this paper.

ACKNOWLEDGMENTS

Thanks are due to Mulan Dai and reviewers for helpful comments on the paper. N. Bazghaleh, A. Navarro-Borrell, C. Yang, and A. Esmaeili Taheri were supported by Saskatchewan Pulse Growers within the Agri-Science Pulse Cluster; W. Ellouze was supported by Western Grain Research Fondation; and Luke D. Bainard was supported by the Agri-Science Organic Cluster.

REFERENCES

1. S. E. Smith and D. J. Read, Mycorrhzial Symbiosis, Academic Press, 2008.
2. E. Lumini, V. Bianciotto, P. Jargeat et al., "Presymbiotic growth and sporal morphology are affected in the arbuscular mycorrhizal fungus Gigaspora margarita cured of its endobacteria," Cellular Microbiology, vol. 9, no. 7, pp. 1716–1729, 2007.
3. J. N. Klironomos and M. M. Hart, "Colonization of roots by arbuscular mycorrhizal fungi using different sources of inoculum," Mycorrhiza, vol. 12, no. 4, pp. 181–184, 2002.
4. C. F. Friese and M. F. Allen, "The spread of VA mycorrhizal fungal hyphae in the soil: inoculum types and external hyphal architecture," Mycologia, vol. 83, no. 4, pp. 409–418, 1991.
5. R. M. Augé, "Water relations, drought and vesicular-arbuscular mycorrhizal symbiosis," Mycorrhiza, vol. 11, no. 1, pp. 3–42, 2001.
6. A. C. Gange and H. M. West, "Interactions between arbuscular mycorrhizal fungi and foliar-feeding insects in Plantago lanceolata L.," New Phytologist, vol. 128, no. 1, pp. 79–87, 1994.
7. K. K. Newsham, A. H. Fitter, and A. R. Watkinson, "Arbuscular mycorrhiza protect an annual grass from root pathogenic fungi in the field," Journal of Ecology, vol. 83, no. 6, pp. 991–1000, 1995.
8. J. N. Klironomos, "Variation in plant response to native and exotic arbuscular Mycorrhizal fungi,"Ecology, vol. 84, no. 9, pp. 2292–2301, 2003.
9. J. P. Grime, J. M. L. Mackey, S. H. Hillier, and D. J. Read, "Floristic diversity in a model system using experimental microcosms," Nature, vol. 328, no. 6129, pp. 420–422, 1988.
10. M. G. A. van der Heijden, J. N. Klironomos, M. Ursic et al., "Mycorrhizal fungal diversity determines plant biodiversity, ecosystem variability and productivity," Nature, vol. 396, no. 6706, pp. 69–72, 1998.
11. M. G. A. van der Heijden, R. Streitwolf-Engel, R. Riedl et al., "The mycorrhizal contribution to plant productivity, plant nutrition and soil structure in experimental grassland," New Phytologist, vol. 172, no. 4, pp. 739–752, 2006.

12. S. Bedini, E. Pellegrino, L. Avio et al., "Changes in soil aggregation and glomalin-related soil protein content as affected by the arbuscular mycorrhizal fungal species Glomus mosseae and Glomus intraradices," Soil Biology and Biochemistry, vol. 41, no. 7, pp. 1491–1496, 2009.
13. M. C. Rillig and D. L. Mummey, "Mycorrhizas and soil structure," New Phytologist, vol. 171, no. 1, pp. 41–53, 2006.
14. A. H. Fitter, T. Helgason, and A. Hodge, "Nutritional exchanges in the arbuscular mycorrhizal symbiosis: Implications for sustainable agriculture," Fungal Biology Reviews, vol. 25, no. 1, pp. 68–72, 2011.
15. T. Fraser, A. Nayyar, W. Ellouze et al., "Arbuscular mycorrhiza: where nature and industry meet," inAdvances in Mycorrhizal Science and Technology, D. Khasa, Y. Piché, and A. P. Coughlan, Eds., pp. 71–86, NRC Research Press and CABI Publishing, Ottawa, Canada, 2009.
16. S. Gianinazzi, A. Gollotte, M.-N. Binet, D. van Tuinen, D. Redecker, and D. Wipf, "Agroecology: the key role of arbuscular mycorrhizas in ecosystem services," Mycorrhiza, vol. 20, no. 8, pp. 519–530, 2010.
17. P. Mäder, F. Kaiser, A. Adholeya et al., "Inoculation of root microorganisms for sustainable wheat-rice and wheat-black gram rotations in India," Soil Biology and Biochemistry, vol. 43, no. 3, pp. 609–619, 2011.
18. A. J. Dumbrell, M. Nelson, T. Helgason, C. Dytham, and A. H. Fitter, "Relative roles of niche and neutral processes in structuring a soil microbial community," International Society for Microbial Ecology Journal, vol. 4, no. 3, pp. 337–345, 2010.
19. L. M. Egerton-Warburton, N. C. Johnson, and E. B. Allen, "Mycorrhizal community dynamics following nitrogen fertilization: a cross-site test in five grasslands," Ecological Monographs, vol. 77, no. 4, pp. 527–544, 2007.
20. A. J. Dumbrell, P. D. Ashton, N. Aziz et al., "Distinct seasonal assemblages of arbuscular mycorrhizal fungi revealed by massively parallel pyrosequencing," New Phytologist, vol. 190, no. 3, pp. 794–804, 2011.

21. A. Heinemeyer, K. P. Ridgway, E. J. Edwards, D. G. Benham, J. P. W. Young, and A. H. Fitter, "Impact of soil warming and shading on colonization and community structure of arbuscular mycorrhizal fungi in roots of a native grassland community," Global Change Biology, vol. 10, no. 1, pp. 52–64, 2004.
22. S. N. Kivlin, C. V. Hawkes, and K. K. Treseder, "Global diversity and distribution of arbuscular mycorrhizal fungi," Soil Biology and Biochemistry, vol. 43, no. 11, pp. 2294–2303, 2011.
23. J. A. Entry, P. T. Rygiewicz, L. S. Watrud, and P. K. Donnelly, "Influence of adverse soil conditions on the formation and function of Arbuscular mycorrhizas," Advances in Environmental Research, vol. 7, no. 1, pp. 123–138, 2002.
24. K. M. Jacobson, "Moisture and substrate stability determine VA-mycorrhizal fungal community distribution and structure in an arid grassland," Journal of Arid Environments, vol. 35, no. 1, pp. 59–75, 1997.
25. A. Antoninka, P. B. Reich, and N. C. Johnson, "Seven years of carbon dioxide enrichment, nitrogen fertilization and plant diversity influence arbuscular mycorrhizal fungi in a grassland ecosystem," New Phytologist, vol. 192, no. 1, pp. 200–214, 2011.
26. J. D. Bever, J. B. Morton, J. Antonovics, and P. A. Schultz, "Host-dependent sporulation and species diversity of arbuscular mycorrhizal fungi in a mown grassland," Journal of Ecology, vol. 84, no. 1, pp. 71–82, 1996.
27. D. Johnson, P. J. Vandenkoornhuyse, J. R. Leake et al., "Plant communities affect arbuscular mycorrhizal fungal diversity and community composition in grassland microcosms," New Phytologist, vol. 161, no. 2, pp. 503–515, 2004.
28. Atul-Nayyar, C. Hamel, T. Forge et al., "Arbuscular mycorrhizal fungi and nematodes are involved in negative feedback on a dual culture of alfalfa and Russian wildrye," Applied Soil Ecology, vol. 40, no. 1, pp. 30–36, 2008.
29. V. A. Borowicz, "Do arbuscular mycorrhizal fungi alter plant-pathogen relations?" Ecology, vol. 82, no. 11, pp. 3057–3068, 2001.
30. C. A. Gehring, J. E. Wolf, and T. C. Theimer, "Terrestrial vertebrates promote arbuscular mycorrhizal fungal diversity and inoculum

potential in a rain forest soil," Ecology Letters, vol. 5, no. 4, pp. 540–548, 2002.

31. J. Jansa, A. Mozafar, T. Anken, R. Ruh, I. R. Sanders, and E. Frossard, "Diversity and structure of AMF communities as affected by tillage in a temperate soil," Mycorrhiza, vol. 12, no. 5, pp. 225–234, 2002.
32. Z. I. Troeh and T. E. Loynachan, "Endomycorrhizal fungal survival in continuous corn, soybean, and fallow," Agronomy Journal, vol. 95, no. 1, pp. 224–230, 2003.
33. D. D. Douds Jr. and P. D. Millner, "Biodiversity of arbuscular mycorrhizal fungi in agroecosystems,"Agriculture, Ecosystems and Environment, vol. 74, no. 1–3, pp. 77–93, 1999.
34. A. Jumpponen, J. Trowbridge, K. Mandyam, and L. Johnson, "Nitrogen enrichment causes minimal changes in arbuscular mycorrhizal colonization but shifts community composition—evidence from rDNA data," Biology and Fertility of Soils, vol. 41, no. 4, pp. 217–224, 2005.
35. T. Helgason, T. J. Daniell, R. Husband, A. H. Fitter, and J. P. W. Young, "Ploughing up the wood-wide web?" Nature, vol. 394, no. 6692, p. 431, 1998.
36. I. Hijri, Z. Sýkorová, F. Oehl et al., "Communities of arbuscular mycorrhizal fungi in arable soils are not necessarily low in diversity," Molecular Ecology, vol. 15, no. 8, pp. 2277–2289, 2006.
37. E. Verbruggen, M. G. A. van der Heijden, J. T. Weedon, G. A. Kowalchuk, and W. F. M. Rö-Ling, "Community assembly, species richness and nestedness of arbuscular mycorrhizal fungi in agricultural soils," Molecular Ecology, vol. 21, no. 10, pp. 2341–2353, 2012.
38. D. Wilson, "Endophyte: the evolution of a term, and clarification of its use and definition," Oikos, vol. 73, no. 2, pp. 274–276, 1995.
39. R. D. Reeleder, "Fungal plant pathogens and soil biodiversity," Canadian Journal of Soil Science, vol. 83, no. 3, pp. 331–336, 2003.
40. R. J. Rodriguez, J. F. White Jr., A. E. Arnold, and R. S. Redman, "Fungal endophytes: diversity and functional roles," New Phytologist, vol. 182, no. 2, pp. 314–330, 2009.

41. K. Clay and C. Schardl, "Evolutionary origins and ecological consequences of endophyte symbiosis with grasses," The American Naturalist, vol. 160, supplement 4, pp. S99–S127, 2002.

42. J. M. Whipps and R. D. Lumsden, "Commercial use of fungi as plant disease biological control agents: status and prospects," in Fungi as Biocontrol Agents: Progress, Problems and Potential, T. M. Butt, C. Jackson, and N. Magan, Eds., pp. 9–22, 2001.

43. B. H. Ownley, M. R. Griffin, W. E. Klingeman, K. D. Gwinn, J. K. Moulton, and R. M. Pereira, "Beauveria bassiana: Endophytic colonization and plant disease control," Journal of Invertebrate Pathology, vol. 98, no. 3, pp. 267–270, 2008.

44. R. Lahlali and M. Hijri, "Screening, identification and evaluation of potential biocontrol fungal endophytes against Rhizoctonia solani AG3 on potato plants," FEMS Microbiology Letters, vol. 311, no. 2, pp. 152–159, 2010.

45. G. E. Harman, "Myths and dogmas of biocontrol: changes in perceptions derived from research onTrichoderma harzianum T-22," Plant Disease, vol. 84, no. 4, pp. 377–393, 2000.

46. J. Liu, G. Gilardi, M. Sanna, M. L. Gullino, and A. Garibaldi, "Biocontrol of Fusarium crown and root rot of tomato and growth-promoting effect of bacteria isolated from recycled substrates of soilless crops," Phytopathologia Mediterranea, vol. 49, no. 2, pp. 163–171, 2010.

47. K. L. McLean, J. S. Hunt, A. Stewart, D. Wite, I. J. Porter, and O. Villalta, "Compatibility of aTrichoderma atroviride biocontrol agent with management practices of Allium crops," Crop Protection, vol. 33, pp. 94–100, 2012.

48. H. Moradi, B. Bahramnejad, J. Amini, A. Siosemardeh, and K. Haji-Allahverdipoor, "Suppression of chickpea (Cicer arietinum L.) Fusarium wilt by Bacillus subtillis and Trichoderma harzianum," Plant Omics Journal, vol. 5, no. 2, pp. 68–74, 2012.

49. K. Mandyam and A. Jumpponen, "Seeking the elusive function of the root-colonising dark septate endophytic fungi," Studies in Mycology, vol. 53, pp. 173–189, 2005.

50. A. E. Arnold, L. C. Mejía, D. Kyllo et al., "Fungal endophytes limit pathogen damage in a tropical tree,"Proceedings of the National

Academy of Sciences of the United States of America, vol. 100, no. 26, pp. 15649–15654, 2003.

51. G. E. Harman, C. R. Howell, A. Viterbo, I. Chet, and M. Lorito, "Trichoderma species—opportunistic, avirulent plant symbionts," Nature Reviews Microbiology, vol. 2, no. 1, pp. 43–56, 2004.

52. M. Tucci, M. Ruocco, L. de Masi, M. de Palma, and M. Lorito, "The beneficial effect of Trichoderma spp. on tomato is modulated by the plant genotype," Molecular Plant Pathology, vol. 12, no. 4, pp. 341–354, 2011.

53. N. Shukla, R. P. Awasthi, L. Rawat, and J. Kumar, "Biochemical and physiological responses of rice (Oryza sativa L.) as influenced by Trichoderma harzianum under drought stress," Plant Physiology and Biochemistry, vol. 54, pp. 78–88, 2012.

54. A. L. Khan, M. Hamayun, Y. Kim, S. Kang, and I. Lee, "Ameliorative symbiosis of endophyte (Penicillium funiculosum LHL06) under salt stress elevated plant growth of Glycine max L.," Plant Physiology and Biochemistry, vol. 49, no. 8, pp. 852–861, 2011.

55. X. Li, N. Bu, Y. Li, L. Ma, S. Xin, and L. Zhang, "Growth, photosynthesis and antioxidant responses of endophyte infected and non-infected rice under lead stress conditions," Journal of Hazardous Materials, vol. 213-214, pp. 55–61, 2012.

56. G. E. Harman, R. Petzoldt, A. Comis, and J. Chen, "Interactions between Trichoderma harzianum strain T22 and maize inbred line Mo17 and effects of these interactions on diseases caused by Pythiuin ultimum and Colletotrichum graminicola," Phytopathology, vol. 94, no. 2, pp. 147–153, 2004.

57. A. Jumpponen, K. G. Mattson, and J. M. Trappe, "Mycorrhizal functioning of Phialocephala fortinii with Pinus contorta on glacier forefront soil: interactions with soil nitrogen and organic matter," Mycorrhiza, vol. 7, no. 5, pp. 261–265, 1998.

58. K. K. Newsham, "Phialophora graminicola, a dark septate fungus, is a beneficial associate of the grassVulpia ciliata ssp. ambigua," New Phytologist, vol. 144, no. 3, pp. 517–524, 1999.

59. J. R. Barrow and P. Osuna, "Phosphorus solubilization and uptake by dark septate fungi in fourwing saltbush, Atriplex canescens (Pursh) Nutt," Journal of Arid Environments, vol. 51, no. 3, pp. 449–459, 2002.

60. L. R. Gale, T. J. Ward, V. Balmas, and H. C. Kistler, "Population subdivision of Fusarium graminearumsensu stricto in the upper midwestern United States," Phytopathology, vol. 97, no. 11, pp. 1434–1439, 2007.
61. A. C. Gange, E. G. Gange, A. B. Mohammad, and L. Boddy, "Host shifts in fungi caused by climate change?" Fungal Ecology, vol. 4, no. 2, pp. 184–190, 2011.
62. A. C. Gange, E. G. Gange, T. H. Sparks, and L. Boddy, "Rapid and recent changes in fungal fruiting patterns," Science, vol. 316, no. 5821, p. 71, 2007.
63. K. E. Fujimura and K. N. Egger, "Host plant and environment influence community assembly of High Arctic root-associated fungal communities," Fungal Ecology, vol. 5, no. 4, pp. 409–418, 2012.
64. J. W. M. Postma, P. A. Olsson, and U. Falkengren-Grerup, "Root colonisation by arbuscular mycorrhizal, fine endophytic and dark septate fungi across a pH gradient in acid beech forests," Soil Biology and Biochemistry, vol. 39, no. 2, pp. 400–408, 2007.
65. J. T. Blodgett, W. J. Swart, S. V. Louw, and W. J. Weeks, "Soil amendments and watering influence the incidence of endophytic fungi in Amaranthus hybridus in South Africa," Applied Soil Ecology, vol. 35, no. 2, pp. 311–318, 2007.
66. J.J.Sadowsky, E.J.Hanson, andA.M.C.Schilder, "Rootcolonization by ericoid mycorrhizae and dark septate endophytes in organic conventional blueberry fields in Michigan," International Journal of Fruit Science, vol. 12, no. 1–3, pp. 169–187, 2012.
67. S. Zubek, A. M. Stefanowicz, J. Błaszkowski, M. Niklińska, and K. Seidler-Łozykowska, "Arbuscular mycorrhizal fungi and soil microbial communities under contrasting fertilization of three medicinal plants," Applied Soil Ecology, vol. 59, pp. 106–115, 2012.
68. A. E. Taheri, C. Hamel, Y. Gan, and V. Vujanovic, "First report of Fusarium redolens from Saskatchewan and its comparative pathogenicity," Canadian Journal of Plant Pathology, vol. 33, no. 4, pp. 559–564, 2011.
69. B. A. D. Hetrick, G. W. T. Wilson, and T. S. Cox, "Mycorrhizal dependence of modern wheat varieties, landraces, and ancestors,"

Canadian Journal of Botany, vol. 70, no. 10, pp. 2032–2040, 1992.

70. B. A. D. Hetrick, G. W. T. Wilson, and T. S. Cox, "Mycorrhizal dependence of modern wheat cultivars and ancestors: a synthesis," Canadian Journal of Botany, vol. 71, no. 3, pp. 512–518, 1993.

71. Y. Zhu, S. E. Smith, A. R. Barritt, and F. A. Smith, "Phosphorus (P) efficiencies and mycorrhizal responsiveness of old and modern wheat cultivars," Plant and Soil, vol. 237, no. 2, pp. 249–255, 2001.

72. A. G. Nelson, S. Quideau, B. Frick, D. Niziol, J. Clapperton, and D. Spaner, "Spring wheat genotypes differentially alter soil microbial communities and wheat breadmaking quality in organic and conventional systems," Canadian Journal of Plant Science, vol. 91, no. 3, pp. 485–495, 2011.

73. R. Azcón and J. A. Ocampo, "Factors affecting the vesicular-arbuscular infection and mycorrhizal dependency of thirteen wheat cultivars," New Phytologist, vol. 87, no. 4, pp. 677–685, 1981.

74. W. Ellouze, H. Yong, C. Hamel, H. Wang, K. Hanson, and A. K. Singh, "Arbuscular mycorrhiza interaction with historical and modern wheat genotypes," Canadian Journal of Plant Science, vol. 92, no. 3, pp. 593–610, 2012.

75. B. A. Hetrick, G. W. Wilson, B. S. Gill, and T. S. Cox, "Chromosome location of mycorrhizal responsive genes in wheat," Canadian Journal of Botany, vol. 73, no. 6, pp. 891–897, 1995.

76. Y. Kapulnik and U. Kushnir, "Growth dependency of wild, primitive and modern cultivated wheat lines on vesicular-arbuscular mycorrhiza fungi," Euphytica, vol. 56, no. 1, pp. 27–36, 1991.

77. A. P. Kirk, M. H. Entz, S. L. Fox, and M. Tenuta, "Mycorrhizal colonization, P uptake and yield of older and modern wheats under organic management," Canadian Journal of Plant Science, vol. 91, no. 4, pp. 663–667, 2011.

78. G. G. B. Manske, "Genetical analysis of the efficiency of VA mycorrhiza with spring wheat," Agriculture, Ecosystems and Environment, vol. 29, no. 1–4, pp. 273–280, 1990.

79. A. K. Singh, C. Hamel, R. M. DePauw, and R. E. Knox, "Genetic variability in arbuscular mycorrhizal fungi compatibility supports the selection of durum wheat genotypes for enhancing soil ecological services and cropping systems in Canada," Canadian Journal of Microbiology, vol. 58, no. 3, pp. 293–302, 2012.

80. R. Singh, R. K. Behl, K. P. Singh, P. Jain, and N. Narula, "Performance and gene effects for wheat yield under inoculation of arbuscular mycorrhiza fungi and Azotobacter chroococcum," Plant, Soil and Environment, vol. 50, no. 9, pp. 409–415, 2004.

81. L. J. C. Xavier and J. J. Germida, "Response of spring wheat cultivars to Glomus clarum NT4 in a P-deficient soil containing arbuscular mycorrhizal fungi," Canadian Journal of Soil Science, vol. 78, no. 3, pp. 481–484, 1998.

82. J. L. Young, E. A. Davis, and S. L. Rose, "Endomycorrhizal fungi in breeder wheats and triticale cultivars field-grown on fertile soil," Agronomy Journal, vol. 77, no. 2, pp. 219–224, 1985.

83. J. B. Baon, S. E. Smith, and A. M. Alston, "Mycorrhizal responses of barley cultivars differing in P efficiency," Plant and Soil, vol. 157, no. 1, pp. 97–105, 1993.

84. R. Koide, M. Li, J. Lewis, and C. Irby, "Role of mycorrhizal infection in the growth and reproduction of wild vs. cultivated plants I. Wild vs. cultivated oats," Oecologia, vol. 77, no. 4, pp. 537–543, 1988.

85. G. H. An, S. Kobayashi, H. Enoki et al., "How does arbuscular mycorrhizal colonization vary with host plant genotype? An example based on maize (Zea mays) germplasms," Plant and Soil, vol. 327, no. 1, pp. 441–453, 2010.

86. T. E. Cheeke, T. N. Rosenstiel, and M. B. Cruzan, "Evidence of reduced arbuscular mycorrhizal fungal colonization in multiple lines of Bt maize," American Journal of Botany, vol. 99, no. 4, pp. 700–707, 2012.

87. S. M. Kaeppler, J. L. Parke, S. M. Mueller, L. Senior, C. Stuber, and W. F. Tracy, "Variation among maize inbred lines and detection of quantitative trait loci for growth at low phosphorus and responsiveness to arbuscular mycorrhizal fungi," Crop Science, vol. 40, no. 2, pp. 358–364, 2000.

88. S. Khalil, T. E. Loynachan, and M. Ali Tabatabai, "Mycorrhizal dependency and nutrient uptake by improved and unimproved corn and soybean cultivars," Agronomy Journal, vol. 86, no. 6, pp. 949–958, 1994.
89. C. A. Oliveira, N. M. H. Sá, E. A. Gomes et al., "Assessment of the mycorrhizal community in the rhizosphere of maize (Zea mays L.) genotypes contrasting for phosphorus efficiency in the acid savannas of Brazil using denaturing gradient gel electrophoresis (DGGE)," Applied Soil Ecology, vol. 41, no. 3, pp. 249–258, 2009.
90. C. Picard, E. Baruffa, and M. Bosco, "Enrichment and diversity of plant-probiotic microorganisms in the rhizosphere of hybrid maize during four growth cycles," Soil Biology & Biochemistry, vol. 40, no. 1, pp. 106–115, 2008.
91. S. S. Dhillion, "Host-endophyte specificity of vesicular-arbuscular mycorrhizal colonization of Oryza sativa l. at the pre-transplant stage in low or high phosphorus soil," Soil Biology and Biochemistry, vol. 24, no. 5, pp. 405–411, 1992.
92. X. Gao, T. W. Kuyper, C. Zou, F. Zhang, and E. Hoffland, "Mycorrhizal responsiveness of aerobic rice genotypes is negatively correlated with their zinc uptake when nonmycorrhizal," Plant and Soil, vol. 290, no. 1-2, pp. 283–291, 2007.
93. H. Nwoko and N. Sanginga, "Dependence of promiscuous soybean and herbaceous legumes on arbuscular mycorrhizal fungi and their response to bradyrhizobial inoculation in low P soils," Applied Soil Ecology, vol. 13, no. 3, pp. 251–258, 1999.
94. G. A. Galván, T. W. Kuyper, K. Burger et al., "Genetic analysis of the interaction between Allium species and arbuscular mycorrhizal fungi," Theoretical and Applied Genetics, vol. 122, no. 5, pp. 947–960, 2011.
95. K. Tawaraya, K. Tokairin, and T. Wagatsuma, "Dependence of Allium fistulosum cultivars on the arbuscular mycorrhizal fungus, Glomus fasciculatum," Applied Soil Ecology, vol. 17, no. 2, pp. 119–124, 2001.
96. D. R. Bryla and R. T. Koide, "Role of mycorrhizal infection in the growth and reproduction of wild vs. cultivated plants—II. Eight wild accessions and two cultivars of Lycopersicon esculentum Mill,"Oecologia, vol. 84, no. 1, pp. 82–92, 1990.

97. O. A. Quilambo, I. Weissenhorn, H. Doddema, P. J. C. Kuiper, and I. Stulen, "Arbuscular mycorrhizal inoculation of peanut in low-fertile tropical soil. I. Host-fungus compatibility," Journal of Plant Nutrition, vol. 28, no. 9, pp. 1633–1644, 2005.
98. R. G. Linderman and E. A. Davis, "Varied response of marigold (Tagetes spp.) genotypes to inoculation with different arbuscular mycorrhizal fungi," Scientia Horticulturae, vol. 99, no. 1, pp. 67–78, 2004.
99. S. Sensoy, S. Demir, O. Turkmen, C. Erdinc, and O. B. Savur, "Responses of some different pepper (Capsicum annuum L.) genotypes to inoculation with two different arbuscular mycorrhizal fungi,"Scientia Horticulturae, vol. 113, no. 1, pp. 92–95, 2007.
100. D. Marshall, B. Tunali, and L. R. Nelson, "Occurrence of fungal endophytes in species of wild Triticum,"Crop Science, vol. 39, no. 5, pp. 1507–1512, 1999.
101. M. Fiers, V. Edel-Hermann, C. Chatot, Y. Le Hingrat, C. Alabouvette, and C. Steinberg, "Potato soil-borne diseases. A review," Agronomy for Sustainable Development, vol. 32, no. 1, pp. 93–132, 2012.
102. P. Garbeva, J. A. van Veen, and J. D. van Elsas, "Microbial diversity in soil: selection of microbial populations by plant and soil type and implications for disease suppressiveness," Annual Review of Phytopathology, vol. 42, no. 1, pp. 243–270, 2004.
103. Y. T. Gan, P. R. Miller, B. G. McConkey, R. P. Zentner, F. C. Stevenson, and C. L. McDonald, "Influence of diverse cropping sequences on durum wheat yield and protein in the semiarid northern Great Plains,"Agronomy Journal, vol. 95, no. 2, pp. 245–252, 2003.
104. V. Vilich, "Crop rotation with pure stands and mixtures of barley and wheat to control stem and root rot diseases," Crop Protection, vol. 12, no. 5, pp. 373–379, 1993.
105. A. G. Nelson, A. Pswarayi, S. Quideau, B. Frick, and D. Spaner, "Yield and weed suppression of crop mixtures in organic and conventional systems of the western Canadian prairie," Agronomy Journal, vol. 104, no. 3, pp. 756–762, 2012.
106. R. E. Blackshaw, L. J. Molnar, and J. R. Moyer, "Suitability of legume cover crop-winter wheat intercrops on the semi-arid

Canadian prairies," Canadian Journal of Plant Science, vol. 90, no. 4, pp. 479–488, 2010.

107. R. G. Smith, K. L. Gross, and G. P. Robertson, "Effects of crop diversity on agroecosystem function: crop yield response," Ecosystems, vol. 11, no. 3, pp. 355–366, 2008.

108. M. Dos Reis Martins, D. A. Angers, and J. E. Corá, "Carbohydrate composition and water-stable aggregation of an oxisol as affected by crop sequence under no-till," Soil Science Society of America Journal, vol. 76, no. 2, pp. 475–484, 2012.

109. L. P. Kiær, I. M. Skovgaard, and H. Østergård, "Grain yield increase in cereal variety mixtures: a meta-analysis of field trials," Field Crops Research, vol. 114, no. 3, pp. 361–373, 2009.

110. R. P. Larkin, T. S. Griffin, and C. W. Honeycutt, "Rotation and cover crop effects on soilborne potato diseases, tuber yield, and soil microbial communities," Plant Disease, vol. 94, no. 12, pp. 1491–1502, 2010.

111. K. M. Harinikumar and D. J. Bagyaraj, "Effect of crop rotation on native vesicular arbuscular mycorrhizal propagules in soil," Plant and Soil, vol. 110, no. 1, pp. 77–80, 1988.

112. A. Fließbach, M. Messmer, B. Nietlispach, V. Infante, and P. Mäder, "Effects of conventionally bred andBacillus thuringiensis (Bt) maize varieties on soil microbial biomass and activity," Biology and Fertility of Soils, vol. 48, no. 3, pp. 315–324, 2012.

113. M. C. Zabaloy, E. Gómez, J. L. Garland, and M. A. Gómez, "Assessment of microbial community function and structure in soil microcosms exposed to glyphosate," Applied Soil Ecology, vol. 61, pp. 333–339, 2012.

114. Y. J. Chun, H. Kim, K. W. Park et al., "Two-year field study shows little evidence that PPO-transgenic rice affects the structure of soil microbial communities," Biology and Fertility of Soils, vol. 48, no. 4, pp. 453–461, 2012.

115. S. Gschwendtner, J. Esperschütz, F. Buegger et al., "Effects of genetically modified starch metabolism in potato plants on photosynthate fluxes into the rhizosphere and on microbial degraders of root exudates," FEMS Microbiology Ecology, vol. 76, no. 3, pp. 564–575, 2011.

116. H. Lu, W. Wu, Y. Chen, X. Zhang, M. Devare, and J. E. Thies, "Decomposition of Bt transgenic rice residues and response of soil microbial community in rapeseed-rice cropping system," Plant and Soil, vol. 336, no. 1, pp. 279–290, 2010.

117. C.-H. Lin and T.-M. Pan, "PCR-Denaturing gradient gel electrophoresis analysis to assess the effects of a genetically modified cucumber mosaic virus-resistant tomato plant on soil microbial communities,"Applied and Environmental Microbiology, vol. 76, no. 10, pp. 3370–3373, 2010.

118. M. C. Zabaloy, G. P. Zanini, V. Bianchinotti, M. A. Gomez, and J. L. Garland, "Herbicides in the soil environment: linkage between bioavailability and microbial ecology," in Herbicides, Theory and Applications, S. A. L. Soloneski and L. Marcelo, Eds., pp. 161–192, InTech, 2011.

119. M. Sheng, C. Hamel, and M. R. Fernandez, "Cropping practices modulate the impact of glyphosate on arbuscular mycorrhizal fungi and rhizosphere bacteria in agroecosystems of the semiarid prairie,"Canadian Journal of Microbiology, vol. 58, no. 8, pp. 990–1001, 2012.

120. N. Z. Lupwayi, S. A. Brandt, K. N. Harker, J. T. O›Donovan, G. W. Clayton, and T. K. Turkington, "Contrasting soil microbial responses to fertilizers and herbicides in a canola-barley rotation," Soil Biology and Biochemistry, vol. 42, no. 11, pp. 1997–2004, 2010.

121. B. Muñoz-Leoz, C. Garbisu, I. Antigüedad, and E. Ruiz-Romera, "Fertilization can modify the non-target effects of pesticides on soil microbial communities," Soil Biology and Biochemistry, vol. 48, pp. 125–134, 2012.

122. Z. A. Siddiqui and K. Futai, "Biocontrol of Meloidogyne incognita on tomato using antagonistic fungi, plant-growth-promoting rhizobacteria and cattle manure," Pest Management Science, vol. 65, no. 9, pp. 943–948, 2009.

123. R. K. Verma and I. D. Arya, "Effect of arbuscular mycorrhizal fungal isolates and organic manure on growth and mycorrhization of micropropagated Dendrocalamus asper plantlets and on spore production in their rhizosphere," Mycorrhiza, vol. 8, no. 2, pp. 113–116, 1998.

124. Ö. Üstüner, S. Wininger, V. Gadkar et al., "Evaluation of different compost amendments with AM fungal inoculum for optimal growth of chives," Compost Science and Utilization, vol. 17, no. 4, pp. 257–265, 2009.
125. F. Oehl, E. Sieverding, K. Ineichen, P. Mäder, T. Boller, and A. Wiemken, "Impact of land use intensity on the species diversity of arbuscular mycorrhizal fungi in agroecosystems of Central Europe," Applied and Environmental Microbiology, vol. 69, no. 5, pp. 2816–2824, 2003.
126. C. Grant and R. Wu, "Enhanced-efficiency fertilizers for use on the Canadian Prairies," Crop Management, vol. 7, no. 1, 2008.
127. V. O. Biederbeck, D. Curtin, O. T. Bouman, C. A. Campbell, and H. Ukrainetz, "Soil microbial and biochemical properties after ten years of fertilization with urea and anhydrous ammonia," Canadian Journal of Soil Science, vol. 76, no. 1, pp. 7–14, 1996.
128. I. Brito, M. J. Goss, M. de Carvalho, O. Chatagnier, and D. van Tuinen, "Impact of tillage system on arbuscular mycorrhiza fungal communities in the soil under Mediterranean conditions," Soil and Tillage Research, vol. 121, pp. 63–67, 2012.
129. J. K. L. Kock and A. Botha, "Fatty acids in fungal taxonomy," in Chemical Fungal Taxonomy, J. C. Frisvad, P. D. Bridge, and D. K. Arora, Eds., pp. 212–246, Marcel Dekker, New York, NY, USA, 1998.
130. Z. Kabir, "Tillage or no-tillage: impact on mycorrhizae," Canadian Journal of Plant Science, vol. 85, no. 1, pp. 23–29, 2005.
131. P. Sarkar, E. Bosneaga, and M. Auer, "Plant cell walls throughout evolution: towards a molecular understanding of their design principles," Journal of Experimental Botany, vol. 60, no. 13, pp. 3615–3635, 2009.
132. G. Berg and K. Smalla, "Plant species and soil type cooperatively shape the structure and function of microbial communities in the rhizosphere," FEMS Microbiology Ecology, vol. 68, no. 1, pp. 1–13, 2009.
133. W. Ellouze, H. Chantal, S. Bouzid, and M. St-Arnaud, "Mycorrhizosphere interactions mediated through rhizodepositions and arbuscular mycorrhizal hyphodeposition and their application in sustainable agriculture," in Mycorrhizal Fungi: Soil, Agriculture and Environmental Implications, S. M.

Fulton, Ed., pp. 133–152, Nova Science Publishers, Hauppauge, NY, USA, 2011.

134. H. Lambers, C. Mougel, B. Jaillard, and P. Hinsinger, "Plant-microbe-soil interactions in the rhizosphere: an evolutionary perspective," Plant and Soil, vol. 321, no. 1-2, pp. 83–115, 2009.

135. F. Wichern, E. Eberhardt, J. Mayer, R. G. Joergensen, and T. Müller, "Nitrogen rhizodeposition in agricultural crops: methods, estimates and future prospects," Soil Biology and Biochemistry, vol. 40, no. 1, pp. 30–48, 2008.

136. F. D. Dakora and D. A. Phillips, "Root exudates as mediators of mineral acquisition in low-nutrient environments," Plant and Soil, vol. 245, no. 1, pp. 35–47, 2002.

137. A. Sugiyama and K. Yazaki, "Root exudates of legume plants and their involvement in interactions with soil microbes," in Secretions and Exudates in Biological Systems, J. M. Vivanco and F. Baluska, Eds., pp. 27–48, Springer, Berlin, Germany, 2012.

138. L. L. M. Fries, R. S. Pacovsky, G. R. Safir, and J. O. Siqueira, "Plant growth and arbuscular mycorrhizal fungal colonization affected by exogenously applied phenolic compounds," Journal of Chemical Ecology, vol. 23, no. 7, pp. 1755–1767, 1997.

139. H. Vierheilig, S. Steinkellner, T. Khaosaad, and J. M. Garcia-Garrido, "The biocontrol effect of mycorrhization on soil-borne fungal pathogens and the autoregulation of the AM symbiosis: one mechanism, two effects?" in Mycorrhiza, A. Varma, Ed., pp. 307–320, Springer, 2008.

140. R. Bari and J. D. G. Jones, "Role of plant hormones in plant defence responses," Plant Molecular Biology, vol. 69, no. 4, pp. 473–488, 2009.

141. E. Dor, D. M. Joel, Y. Kapulnik, H. Koltai, and J. Hershenhorn, "The synthetic strigolactone GR24 influences the growth pattern of phytopathogenic fungi," Planta, vol. 234, no. 2, pp. 419–427, 2011.

142. T. Kretzschmar, W. Kohlen, J. Sasse et al., "A petunia ABC protein controls strigolactone-dependent symbiotic signalling and branching," Nature, vol. 483, no. 7389, pp. 341–344, 2012.

143. S. Steinkellner, V. Lendzemo, I. Langer et al., "Flavonoids and strigolactones in root exudates as signals in symbiotic and

pathogenic plant-fungus interactions," Molecules, vol. 12, no. 7, pp. 1290–1306, 2007.

144. S. Horii, A. Matsumura, M. Kuramoto, and T. Ishii, "Tryptophan dimer produced by water-stressed bahia grass is an attractant for Gigaspora margarita and Glomus caledonium," World Journal of Microbiology and Biotechnology, vol. 25, no. 7, pp. 1207–1215, 2009.

145. B. Fritig, T. Heitz, and M. Legrand, "Antimicrobial proteins in induced plant defense," Current Opinion in Immunology, vol. 10, no. 1, pp. 16–22, 1998.

146. A. Turrini, C. Sbrana, L. Pitto et al., "The antifungal Dm-AMP1 protein from Dahlia merckii expressed in Solanum melongena is released in root exudates and differentially affects pathogenic fungi and mycorrhizal symbiosis," New Phytologist, vol. 163, no. 2, pp. 393–403, 2004.

147. W. Ellouze, C. Hamel, A. F. Cruz et al., "Phytochemicals and spore germination: at the root of AMF host preference?" Applied Soil Ecology, vol. 60, pp. 98–104, 2012.

148. M. A. Birkett, T. J. A. Bruce, J. L. Martin, L. E. Smart, J. Oakley, and L. J. Wadhams, "Responses of female orange wheat blossom midge, Sitodiplosis mosellana, to wheat panicle volatiles," Journal of Chemical Ecology, vol. 30, no. 7, pp. 1319–1328, 2004.

149. A. F. Cruz, C. Hamel, C. Yang et al., "Phytochemicals to suppress fusarium head blight in wheat-chickpea rotation," Phytochemistry, vol. 78, pp. 72–80, 2012.

150. K. Yoneyama, X. Xie, H. Sekimoto et al., "Strigolactones, host recognition signals for root parasitic plants and arbuscular mycorrhizal fungi, from Fabaceae plants," New Phytologist, vol. 179, no. 2, pp. 484–494, 2008.

151. M. F. Allen and M. G. Boosalis, "Effects of two species of VA mycorrhizal fungi on drought tolerance of winter wheat," New Phytologist, vol. 93, no. 1, pp. 67–76, 1983.

152. M. Vivekanandan and P. E. Fixen, "Cropping systems effects on mycorrhizal colonization, early growth, and phosphorus uptake of corn," Soil Science Society of America Journal, vol. 55, no. 1, pp. 136–140, 1991.

153. K. O. Reinhart, G. W. T. Wilson, and M. J. Rinella, "Predicting plant responses to mycorrhizae: integrating evolutionary history and plant traits," Ecology Letters, vol. 15, no. 7, pp. 689–695, 2012.

154. Z. Kabir, I. P. O›Halloran, and C. Hamel, "Overwinter survival of arbuscular mycorrhizal hyphae is favored by attachment to roots but diminished by disturbance," Mycorrhiza, vol. 7, no. 4, pp. 197–200, 1997.

155. W. Ellouze, C. Hamel, V. Vujanovic, Y. Gan, S. Bouzid, and M. St-Arnaud, "Chickpea genotypes shape the soil microbiome and affect the establishment of the subsequent durum wheat crop in the semiarid North American Great Plains," Soil Biology & Biochemistry, vol. 63, pp. 129–141, 2013.

156. J. C. Sillero, A. M. Villegas-Fernández, J. Thomas et al., "Faba bean breeding for disease resistance,"Field Crops Research, vol. 115, no. 3, pp. 297–307, 2010.

157. K. Egle, "Improved phosphorus efficiency of three new wheat genotypes from CIMMYT in comparison with an older Mexican variety," Journal of Plant Nutrition and Soil Science, vol. 162, no. 3, pp. 353–358, 1999.

158. A. Lehmann, E. K. Barto, J. R. Powell, and M. C. Rillig, "Mycorrhizal responsiveness trends in annual crop plants and their wild relatives-a meta-analysis on studies from 1981 to 2010," Plant and Soil, vol. 355, no. 1-2, pp. 231–250, 2012.

159. M. R. Finckh, "Integration of breeding and technology into diversification strategies for disease control in modern agriculture," in Sustainable Disease Management in a European Context, D. B. Collinge, L. Munk, and B. M. Cooke, Eds., pp. 399–409, Springer, Amsterdam, The Netherlands, 2008.

160. B. Pivato, S. Mazurier, P. Lemanceau et al., "Medicago species affect the community composition of arbuscular mycorrhizal fungi associated with roots," New Phytologist, vol. 176, no. 1, pp. 197–210, 2007.·

161. A. J. Bennett, G. D. Bending, D. Chandler, S. Hilton, and P. Mills, "Meeting the demand for crop production: the challenge of yield decline in crops grown in short rotations," Biological Reviews, vol. 87, no. 1, pp. 52–71, 2012.

162. W. Ellouze, K. Hanson, A. Nayyar, J. Perez, and C. Hamel, "Intertwined existence: the life of plant symbiotic fungi in agricultural soils," in Mycorrhiza, A. Varma, Ed., pp. 507–528, Springer, Berlin, Germany, 2008.
163. E. Bernard, R. P. Larkin, S. Tavantzis et al., "Compost, rapeseed rotation, and biocontrol agents significantly impact soil microbial communities in organic and conventional potato production systems," Applied Soil Ecology, vol. 52, no. 1, pp. 29–41, 2012.
164. C. G. Castillo, R. Rubio, J. L. Rouanet, and F. Borie, "Early effects of tillage and crop rotation on arbuscular mycorrhizal fungal propagules in an Ultisol," Biology and Fertility of Soils, vol. 43, no. 1, pp. 83–92, 2006.
165. A. Nayyar, C. Hamel, G. Lafond, B. D. Gossen, K. Hanson, and J. Germida, "Soil microbial quality associated with yield reduction in continuous-pea," Applied Soil Ecology, vol. 43, no. 1, pp. 115–121, 2009.
166. R. P. Larkin and C. W. Honeycutt, "Effects of different 3-year cropping systems on soil microbial communities and rhizoctonia diseases of potato," Phytopathology, vol. 96, no. 1, pp. 68–79, 2006.
167. H. Swer, M. S. Dkhar, and H. Kayang, "Fungal population and diversity in organically amended agricultural soils of Meghalaya, India," Journal of Organic Systems, vol. 6, no. 2, pp. 3–12, 2011.
168. U. Merz and R. E. Falloon, "Review: powdery scab of potato—increased knowledge of pathogen biology and disease epidemiology for effective disease management," Potato Research, vol. 52, no. 1, pp. 17–37, 2009.
169. Y. Gan, C. Liang, C. Hamel, H. Cutforth, and H. Wang, "Strategies for reducing the carbon footprint of field crops for semiarid areas. A review," Agronomy for Sustainable Development, vol. 31, no. 4, pp. 643–656, 2011.
170. M. E. Gavito and M. H. Miller, "Changes in mycorrhiza development in maize indeed by crop management practices," Plant and Soil, vol. 198, no. 2, pp. 185–192, 1998.
171. E. Njeru, L. Avio, C. Sbrana et al., "First evidence for a major cover crop effect on arbuscular mycorrhizal fungi and organic maize growth," Agronomy for Sustainable Development, 2013.

172. P. Bednarek and A. Osbourn, "Plant-microbe interactions: chemical diversity in plant defense," Science, vol. 324, no. 5928, pp. 746–748, 2009.
173. A. Scalbert and G. Williamson, "Dietary intake and bioavailability of polyphenols," The Journal of Nutrition, vol. 130, no. 8, pp. 2073S–2085S, 2000.
174. J. B. Morris and J. T. Walker, "Non-traditional legumes as potential soil amendments for nematode control," Journal of Nematology, vol. 34, no. 4, pp. 358–361, 2002.
175. M. F. Cohen, H. Yamasaki, and M. Mazzola, "Brassica napus seed meal soil amendment modifies microbial community structure, nitric oxide production and incidence of Rhizoctonia root rot," Soil Biology and Biochemistry, vol. 37, no. 7, pp. 1215–1227, 2005.
176. J. Bridge, "Nematode management in sustainable and subsistence agriculture," Annual Review of Phytopathology, vol. 34, pp. 201–225, 1996.
177. A. O. Adesemoye and J. W. Kloepper, "Plant-microbes interactions in enhanced fertilizer-use efficiency," Applied Microbiology and Biotechnology, vol. 85, no. 1, pp. 1–12, 2009.
178. A. O. Adesemoye, H. A. Torbert, and J. W. Kloepper, "Enhanced plant nutrient use efficiency with PGPR and AMF in an integrated nutrient management system," Canadian Journal of Microbiology, vol. 54, no. 10, pp. 876–886, 2008.
179. R. J. Manlay, C. Feller, and M. J. Swift, "Historical evolution of soil organic matter concepts and their relationships with the fertility and sustainability of cropping systems," Agriculture, Ecosystems and Environment, vol. 119, no. 3-4, pp. 217–233, 2007.
180. M. H. Beare, P. F. Hendrix, and D. C. Coleman, "Water-stable aggregates and organic matter fractions in conventional- and no-tillage soils," Soil Science Society of America Journal, vol. 58, no. 3, pp. 777–786, 1994.
181. R. Raghuwanshi and R. S. Upadhyay, "Performance of vesicular-arbuscular mycorrhizae in saline-alkali soil in relation to various amendments," World Journal of Microbiology and Biotechnology, vol. 20, no. 1, pp. 1–5, 2004.

182. Statistics-Canada, "Area of commercial fertilizer, herbicides, insecticides and fungicides applied, by province," Census of Agriculture, 1996 to 2006, 2008, http://www.statcan.gc.ca/tables-tableaux/sum-som/l01/cst01/agrc05i-eng.htm.

183. M. Korol, "Fertilizer and pesticide management in Canada," Farm Environmental Management in Canada, vol. 1, no. 3, pp. 1–41, 2004, http://www.statcan.gc.ca/pub/21-021-m/21-021-m2004002-eng.htm.

184. F. Adams, "Crop response to lime in the southern United States," in Soil Acidity and Liming, F. Adams, Ed., pp. 211–265, AS CSSA, SSSA, Madison, Wis, USA, 1984.

185. O. T. Bouman, D. Curtin, C. A. Campbell, V. O. Biederbeck, and H. Ukrainetz, "Soil acidification from long-term use of anhydrous ammonia and urea," Soil Science Society of America Journal, vol. 59, no. 5, pp. 1488–1494, 1995.

186. D. W. McAndrew and S. S. Malhi, "Long-term N fertilization of a solonetzic soil: effects on chemical and biological properties," Soil Biology and Biochemistry, vol. 24, no. 7, pp. 619–623, 1992.

187. S. J. Patterson, S. N. Acharya, J. E. Thomas, A. B. Bertschi, and R. L. Rothwell, "Barley biomass and grain yield and canola seed yield response to land application of wood ash," Agronomy Journal, vol. 96, no. 4, pp. 971–977, 2004.

188. J. Rousk, E. Bååth, P. C. Brookes et al., "Soil bacterial and fungal communities across a pH gradient in an arable soil," The ISME Journal, vol. 4, no. 10, pp. 1340–1351, 2010.

189. M. Wainwright, "A review of the effects of pesticides on microbial activity in soils," Journal of Soil Science, vol. 29, no. 3, pp. 287–298, 1978.

190. E. K. Bünemann, G. D. Schwenke, and L. van Zwieten, "Impact of agricultural inputs on soil organisms—a review," Soil Research, vol. 44, no. 4, pp. 379–406, 2006.

191. S. Hussain, T. Siddique, M. Saleem, M. Arshad, and A. Khalid, "Impact of pesticides on soil microbial diversity, enzymes, and biochemical reactions," in Advances in Agronomy, L. S. Donald, Ed., pp. 159–200, Academic Press, New York, NY, USA, 2009.

192. C. Zhang, X. Liu, F. Dong, J. Xu, Y. Zheng, and J. Li, "Soil microbial communities response to herbicide 2,4-dichlorophenoxyacetic acid butyl ester," European Journal of Soil Biology, vol. 46, no. 2, pp. 175–180, 2010.

Chapter 8

The Relationship between Soil PH and Selected Soil Properties in 48 Years Logged-over Forest

Mohd-Aizat, A., Mohamad-Roslan, M.K.[1], Wan Nor Azmin Sulaiman[2], and Daljit Singh Karam[3]

[1]Department of Forest Production, Faculty of Forestry, Universiti Putra Malaysia, 43400 UPM Serdang, Selangor, Malaysia

[2]Department of Environmental Sciences, Universiti Putra Malaysia 43400 Serdang, Selangor Darul Ehsan, Malaysia

[3]Department of Land Management, Faculty of Agriculture, Universiti Putra Malaysia, 43400 UPM Serdang, Selangor, Malaysia

ABSTRACT

Selectively logged occur at Ayer Hitam Forest Reserve starting from 1936 to 1965. The geology of this forest is made of igneous rocks with granite which is a common acid rock as its main component that may probably affect the soil properties. The relationship between soil pH with selected soil biological and chemical properties were investigated. Results showed that, the selected soil properties related mostly in the form of power function where the best simple linear regression model for soil microbial biomass C, soil microbial biomass N and total Fe with soil pH were (ln (Biomass C) = 4.534 ln (Soil pH), with R^2= 0.996) (ln (Biomass N) = 3.035 ln (Soil pH), R^2= 0.994) and (ln (Iron) = 4.121 ln (Soil pH),R^2=0.996 respectively. However, it has a negative relationship with soil electrical conductivity as the best in the equation was (ln (Soil EC) = -1.037 ln (Soil pH), R^2=0.973). The amount of copper content in the soil is related positively in the form of logarithmic function with soil pH (Copper = 0.877 in (Soil pH), R^2=0.621).This work has establish clear relationship between soil pH with selected soil properties in acidic soil condition.

INTRODUCTION

Malaysia currently is the leading developed nation in Southeast Asia and one of the fastest developed countries in the world. Many forest areas in Malaysia have been converted to urban development and logging activities for timber production in order to fulfil the increase of human population, need for job security and the demands for food (Potts *et al.,* 2005). In a tropical forest, soils are at acidic conditions for most soil series in the pH range of 4.5 to 5.5 but certain tropical soils are highly weathered which have pH of about 4.0 such as Ultisols and Oxisols (Shamshudin and Fauziah, 2010). Tropical rainforest soils are generally highly weathered soil status due to climate conditions. Soil pH affects all biological, chemical and physical soil properties (Brady and Weil 2002).

Ayer Hitam Forest Reserve was the forest once being selectively logged about 48 years ago from 1936 to 1965 (Paiman and Ramsa, 2007) with housing and industrial area which are also once part of the forest fencing the area. In 48 years period, forest has undergone recovery in term of plant diversity and soil structure compared to new logged over forest. The geology of Ayer Hitam Forest Reserve is made of igneous rocks with granite as its main component (Paiman and Ramsa, 2007). Igneous rocks that which usually composed of granite are generally acidic. Cu is found in a wide range of soil materials such as igneous rocks, sandstones, limestones and shales. Soils that are derived from silica-rich rocks such as granite, organic soils and podzols are low in Fe content. Soil pH is probably at least as important as soil N and C concentrations in influencing the size of soil microbial biomass (Wardle, 1992). Soil electrical conductivity relates directly to salinity (Pathak and Rao, 1998). Salinity usually refers to the presence of soluble salt in the soil. Soil pH may probably affect the solubility of salts. Therefore, this study was undertaken to examine the relationship between soil pH and selected soil properties at Ayer Hitam Forest Reserve Puchong, Selangor.

MATERIAL AND METHODS

The study was conducted at Ayer Hitam Forest Reserve Puchong Selangor which comprises of six compartments; Compartment 1, Compartment 2, Compartment 12, Compartment 13, Compartment 14 and Compartment 16 with space of 1,248 hectares (Figure 1). Eight sampling stations were set up where Station 1 and 2 was located in compartment 13 which is far from human activities while station 3 to 8 especially station 7 and 8 which are in compartment 15 were close to the human activities.

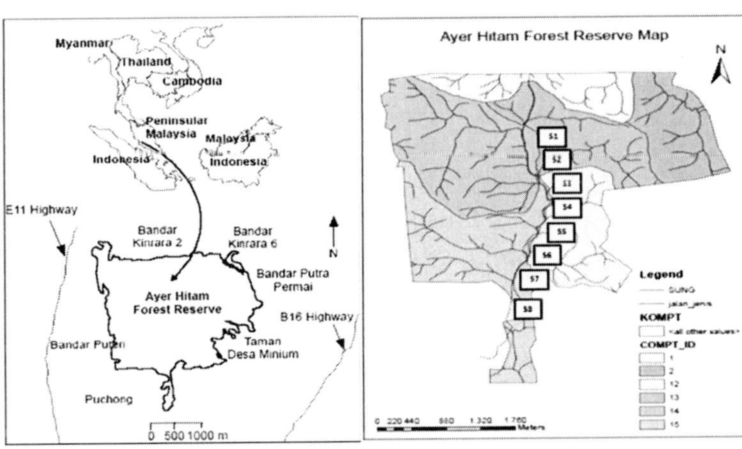

Figure 1: Map location of station at Ayer Hitam Forest Reserve [Note:S1, Station 1;S2, Station 2; S3, Station 3; S4, Station 4; S5, Station 5; S6, Station 6; S7, Station 7; S8, Station 8] Source: ArcGIS / 20 December 2012.

A soil sample that is composed of four sub-samples (4 replicates) to represent the selected soil properties at each sampling station is carried out weekly for the duration of seven weeks. The soil samples were taken randomly with consideration on field condition such as landscape, land morphology, area coverage management technique, vegetation cover and others. The soils were sampled at the depth of 20cm from the soil surface about 5m away from the river to avoid additional sediment from river reserve.

The six soil properties selected were pH, microbial biomass C, microbial biomass N, electrical conductivity, total Fe and total Cu because the changes of these parameter always related with soil pH. Soil pH was determined in 1: 2.5 ratios of soil in distilled water and electrical conductivity was measured using and electrical conductivity meter in 1:1 soil to water ratios. Chloroform fumigation and extraction method were used to determine the microbial biomass based on Anderson and Ingram (1993), cited by Okalebo *et al.* (1993).Concentration of Fe and Cu were determined by the wet digestion method, using the aqua regia solution and analysed by Absorption Spectrophotometer. International standard methods were used for both elements in "aqua regia", a mixture of hydrochloric acid and nitric acid (International Standard ISO 11466, 1995).

RESULTS

Soil pH affects all the physical, biological and chemical soil properties (Brady and Weil, 2002) and the growth of specific organisms, soil microbial biomass, and microbial activity. In this research, it is shown that the range of the soil pH throughout all the stations is from 3.12 to 5.28. Table 1 shows the average soil pH in each sampling station.

Table 1: Average soil pH value for each sampling station

Week	Average Soil pH							
	S1	S2	S3	S4	S5	S6	S7	S8
1	4.72	4.08	4.48	4.66	4.20	4.72	4.02	4.52
2	4.69	4.43	4.56	4.32	4.52	4.71	4.12	4.62
3	4.46	3.95	4.27	4.38	4.01	3.12	3.51	4.02
4	4.46	4.27	4.28	4.78	4.75	4.95	4.36	4.57
5	4.45	4.83	4.54	4.31	4.52	4.62	4.92	4.74
6	5.26	4.52	4.81	4.62	4.71	4.65	4.91	4.96
7	5.28	4.61	4.86	4.66	4.62	4.80	4.71	5.06

Throughout the seven weeks of experiment, the highest soil pH value was recorded at Station 1 while the lowest was at Station 6. However, between weeks, the highest and lowest values were not fixed to those particular stations. There are several factors affecting the overall soil pH such as parent material, climate, living organism, topography, time, native vegetation, crop grown, organic material, precipitation, temperature and human activities. There is little amount of organic matters in Station 1 near former logging road due to low density of plants, causing only little root respiration process and decomposition of organic matter taking place in the area . Because of this, low amount of carbon dioxide from root respiration and decomposition of organic matter dissolve in water. Carroll and Mather (1992) stated that this process releases carbon dioxide in which, when mixes with water in the soil, it will form weak carbonic acid (H_2CO_3). Chemical equation for carbonic acid formation is shown as below:

$$CO_2 + H_2O \rightarrow H_2CO_3 \rightarrow HCO_3 + H^+$$

Due to little amount of organic matter and low density of plants in Station 1, there was low production of carbonic acid in the soil which could cause increment of soil pH level compared to other stations.

According to Drapper and Smith (1981), regression analysis is a statistical technique used to relate the dependent and independent variables. The regression model for relationship between soil pH and selected soil properties are showed in Table 2.

Table 2: Relationship between soil pH and selected soil properties

Variables	**Best Regression Model**	R^2	**Standard error of estimate**	**N**	**Relationship with WQI**
Soil Microbial Biomass C	ln (Biomass C) = 4.534 ln (Soil pH)	0.995	0.466	56	Positive (Power)
Soil Microbial Biomass N	ln (Biomass N) = 3.035 ln (Soil pH)	0.994	0.373	56	Positive (Power)
Soil Electrical Conductivity	ln (Soil EC) = -1.037 ln (Soil pH)	0.973	0.263	56	Negative (Power)
Total Fe	ln (Fe) = 4.121ln (soil pH)	0.996	0.399	56	Positive (Power)
Total Cu	Cu =0.877 ln(Soil pH)	0.621	1.042	56	Positive (Logarithmic)

This research shows that the range of soil microbial biomass C is from 618.11 (µg/g) to 1386.54 (µg/g). The highest concentration of soil microbial biomass C was observed at Station 3 and the lowest of soil microbial biomass C was at Station 4. Soil moisture content is a main control on soil microbial activity and microbial community structure in the forest floor (Wagener and Schimel, 1998). Station 4 has the lowest concentration of soil microbial biomass C as it has the lowest moisture content (percentage of moisture content=9.31%) in the soil than other stations, which hinders the microbes' growth. Although Station 7 having the highest moisture content (percentage of moisture content = 22.45%), Station 7 does not promote growth as it should be. This is due to the fact that it has lower total mean of soil pH than Station 3; low enough to restrain the activities of the microbes. Figure 2 illustrate a graph of the relationship between soil pH and soil microbial biomass C.

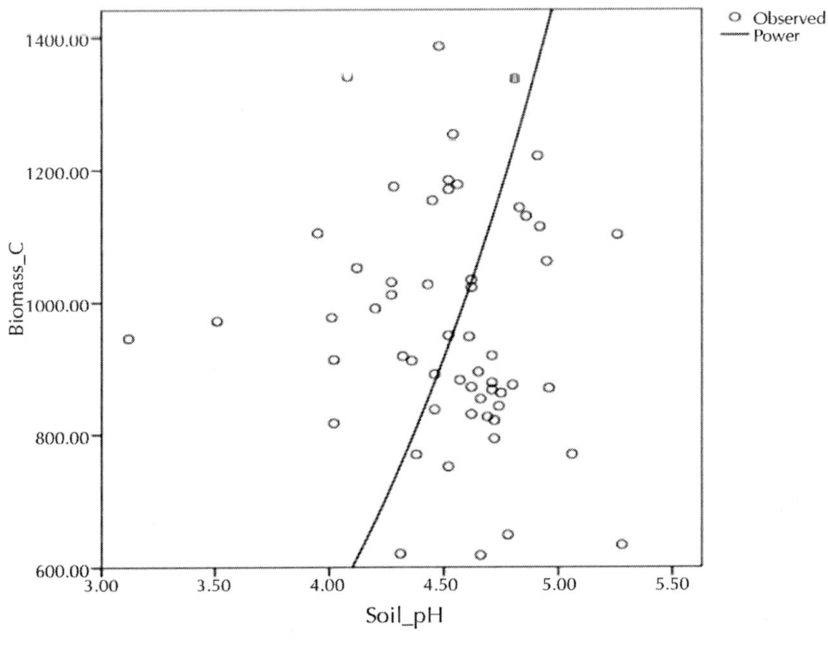

Figure 2: Relationship between soil pH and Biomass C (µg/g).

The concentration of soil microbial biomass C is the greatest at pH 7.00 (Aciego Pietri *et al.*, 2008) and can degrade or immobilize pollutants in the soil. Its mean that in acidic soil, the nearly neutral condition indicates the high content of soil microbial biomass C present at the station. Logah *et al.* (2010) stated that soil microbial biomass C positively related with the soil organic C. Soil organic carbon is the major source of energy for the soil microorganisms (Edward *et al.,* 1999). Organic C reaches its optimum value when the soil pH is in between 5.00 to 6.00 and decreases when lower than 5.00 (Aciego Pietri *et al.* 2008). Soil that rich in organic carbon contains a lot of microorganisms due to high source of energy and thus heighten the concentration of soil microbial biomass C.Decrease in soil pH value will increase the availability of biologically toxic aluminium (Flis *et al.,* 1993). The mechanisms operate in the microbial cells are similar to the mechanisms of those caused by heavy metal. The higher concentration of aluminium in acidic soil may indicate low input of organic C from plants to the soils and will reduce the efficiency of conversion of this C into biomass C. (Aciego Pietri and Brookes, 2008).Soil pH is related positively with soil microbial biomass C in the form of power function and not show in linear relationship because soil microbial biomass C is also influenced by soil conditions, nutrient supply, vegetative cover, soil temperature, light intensity and soil moisture content (Campbell *et al.,* 1999).

Station 4 has the highest soil microbial biomass N which is 143.76 (μg/g) and the lowest is at Station 1 which is 59.65 (μg/g).Station 4 has the highest concentration of soil microbial biomass N because it contains a lot of organic matter from the litter fall due to the denser leafy canopy compares to the Station 1. The decomposition process is reflected by the increase in amount of N and the decrease in C-to-N ratio value of the litter (Gallardo and Merino, 1992). Greater input of N may occur in the soil through higher decomposition rate of litter fall (Singh *et al.*

2010).Greater input of total N tends to increase the content of soil microbial biomass N at this station. Figure 3 illustrate a graph of the relationship between soil pH and soil microbial biomass N.

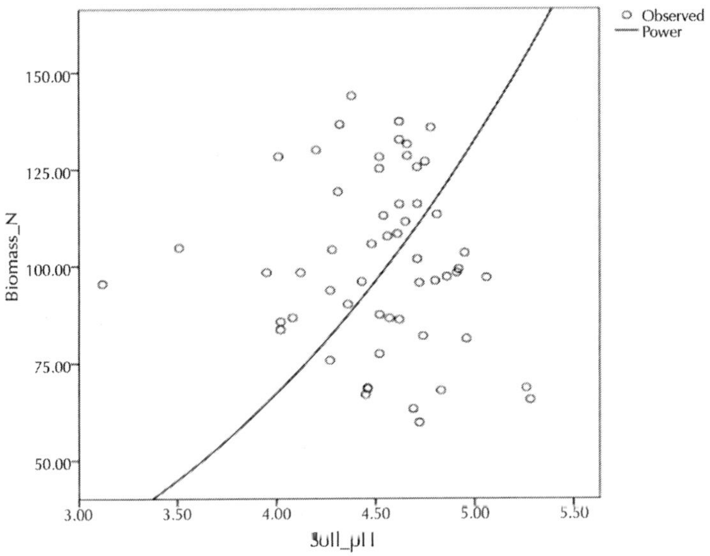

Figure 3: Relationship between Soil pH and Biomass N (μg/g).

Diversity and composition of soil bacterial communities across the entire gradient has strong influence on soil pH. This verifies the positive relationship between bacterial diversity and higher soil pH in the interval pH from 4.00 to 7.00 (Rousk *et al.*, 2010). This implicates that when the soil reaches neutrality, the number of bacterial communities' increases. It is also established that organic C positively relates with total N in the soil (Li *et al.*, 2007). The dynamics of N in mineral soil is closely related to C, because most N exists in organic compounds and heterotrophic microbial biomass, due to consumption of organic C as energy source. This also shows that the microbial biomass N is also positively related with microbial biomass C (Wright *et al.*, 2005). According to Kara and Bolat (2007), there are positive and significant relations between microbial biomass N and total N in the soil. Since organic C reaches optimum value at soil pH of around 5.00 to 6.00 (Logah *et al.*, 2010), it increases total N in the soil and indicates the increase in soil microbial biomass N when soil reaches neutrality. Soil pH and soil microbial biomass N related positively in the form of power function and does not show linear relationship because soil microbial biomass N content is also affected by the vegetative cover, soil conditions, physical and chemical properties, local climatic condition, temperature, moisture content, substrate quality and plant community composition (Li *et al.,* 2012).

Station 4 has the lowest soil electrical conductivity which is 0.12 ds/m and the highest of soil electrical conductivity is at Station 8 which is 0.44 ds/m. There are several factors that affect the soil electrical conductivity such as pore continuity, water content, salinity level, cation exchange capacity, depth and temperature (Geonics, 1980). Station 8 which is close to bitumen-paved road may possibly have high soil electrical conductivity. This occurs due to road salts being the one of the major contributor for nonpoint source pollution from streets (Kramme, 1985). Station 8 which is close to bitumen paved road may possibly have high soil electrical conductivity. Soil moisture also plays an important role in conducting electricity in the soil. The amount of moisture content is directly proportional to the ability to conduct electricity. This effect can be seen at Station 4, where this area has the lowest value of soil electrical conductivity due to its having the lowest amount of moisture content (9.31%). Soil pH is not directly affect the soil electrical conductivity but may probably affect the solubility of salts and soil moisture content which electrical conductivity of the soil. Figure 4 illustrate a graph of the relationship between soil pH and soil electrical conductivity.

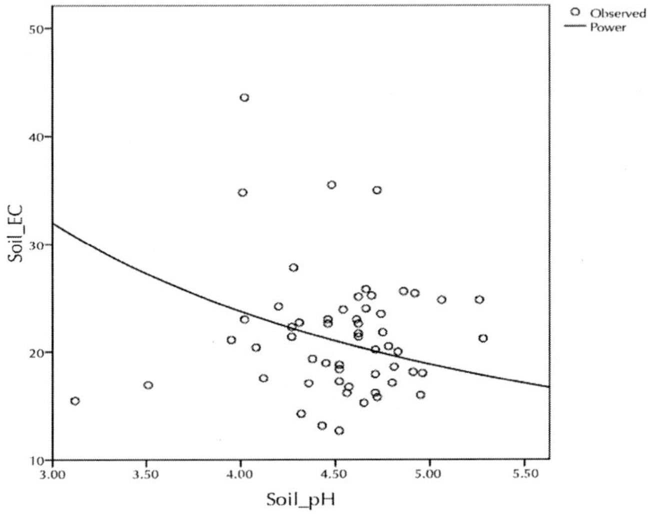

Figure 4: Relationship between Soil pH and Soil EC(dS/m).

Soil electrical conductivity relates directly to salinity (Pathak and Rao, 1998). Salinity usually refers to the presence of soluble salt in the soil. Soil pH may probably affect the solubility of salts and soil moisture content. More alkaline soil will have less amount of soluble salt (Provin *et al.,* 2001). This means that low soil pH value, as found in all stations, should have high soluble salt content and therefore high electrical conductivity. According to Bruckner (2012), lower soil pH indicates larger number of hydrogen ions in the soil. Hydrogen ions can appear in varying amount in the soil environment which can affect the level of electrical conductivity. Higher amount of hydrogen ions in the soil will show a higher rate of electrical conductivity. Hence, low soil pH due to large number of hidrogen ions in the soil may encourage soil electrical conductivity. Soil pH is negatively related with soil electrical conductivity in the form of power function and not in linear relationship because there are several other factors such as soil mineral, porosity, soil texture, soil moisture and soil temperature which also affect soil electrical conductivity in the soil (USDA, 2011).

This research shows that Station 3 has the lowest total Fe in soil which is 246.68 mg/kg and the highest of total Fe in soil is at Station 7 which is 611.87 mg/kg. According to U.S. Environmental Protection Agency (1997), Fe could be directly produced from bitumen-paved road, steel building structure, automobile body rust and moving engine parts. In the context of Station 7 which is close to the bitumen-paved road and educational facilities, steel building structure, automobile body rust and moving engine parts could probably be the reason of high amount of Fe in the area. Figure 5 illustrate a graph of the relationship between soil pH and total Fe.

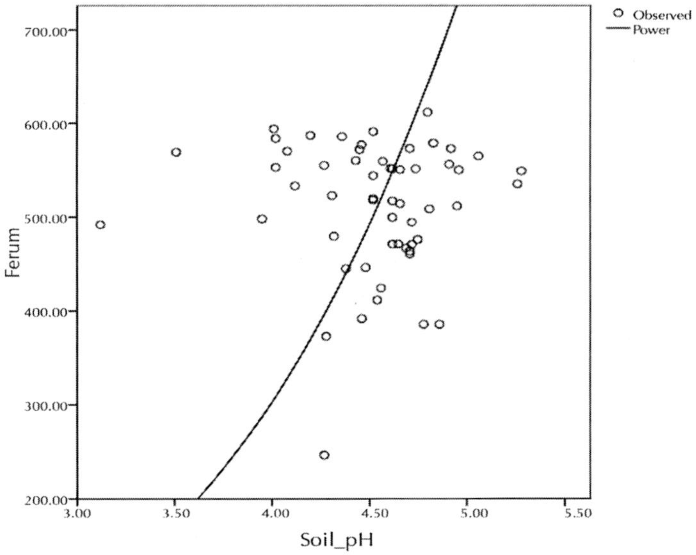

Figure 5: Relationship between soil pH and total Fe (mg/kg).

Total concentration of Fe in the soil also depends on the concentration of organic matter (Schulte, 2004). Organic matter is important in providing iron, sulphur, phosphorus and nitrogen through the decomposition of organic matter by the soil microorganisms. According to Aciego Pietri and Brokes (2008) microbial activities is optimum at pH 7.00. The increase in soil pH to neutral value tends to increase the content of soil microbial biomass and enhance the microbial activity which leads to increase the content of Fe in the soil. The content of Fe has significant correlation with the pH of the soil (Jelic *et al.* 2010). Juzefaciuk *et al.* (1995) found that the highest amount of total Fe is present at the natural soil pH condition. Soil pH relates positively with Fe content in the form of power function and not show in linear relationship because Fe content in the soil is also influenced by the soil conditions (saturated, compacted, poor aerated soil, the presence of other nutrients, leaching and soil erosion (Bill, 2010).

The range of total Cu in the soils throughout all the stations is from 0.084 mg/kg to 4.296 mg/kg. The highest value was recorded at Station 5 and the lowest value was at Station 6. In general, however, these rankings are not particular to the same stations. Soil parameters, likes organic content, pH, iron and manganese oxide and total metal content will affect the distribution of Cu among different soil fractions (Fic and Schroter, 1989). The low concentration of total Cu at Station 6 was probably due to the location of this station at steep area close to the former logging. This tends to create high peak discharge intensity and stormwater volumes during rainfall. This high intensity can cause soil erosion and possibly decrease the total Cu content at Station 6. The significant relationship between water quality index and total Cu in the soil exists when using logarithmic function. Figure 6 illustrate a graph of the relationship between soil pH and total Cu.

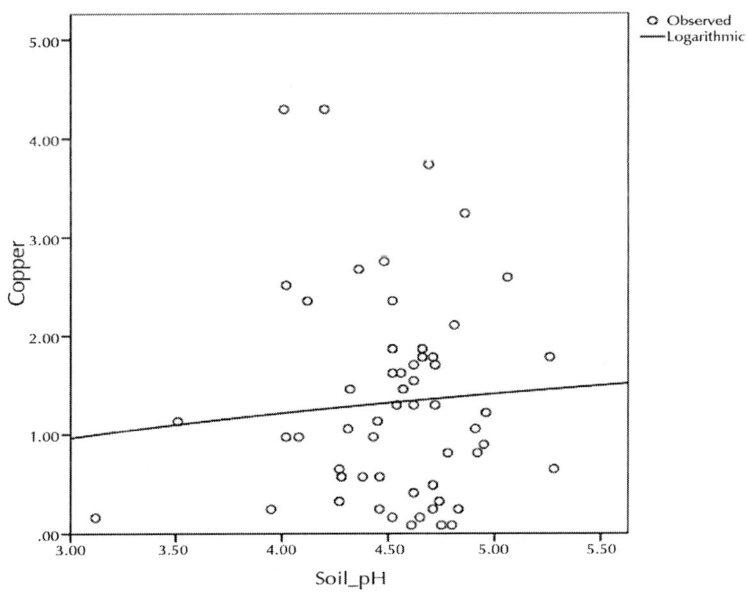

Figure 6: Relationship between Soil pH and Total Cu (mg/kg).

Risti *et al.* (2006) stated that the increase of the total Cu concentration is probably due to higher permeability and increased Cu mobility in the soil. A study by Hossam (2001) found that the mobility of Cu was the highest at pH 6.24. This means that the increase in Cu mobility when soil reaches 6.24 indicates the increment of total Cu concentration in the soil. Total concentration of Cu was significantly correlated with the content of organic matter (Mitsios *et al.* 2005). Soil microbial biomass is a fraction of organic matter in the soil. Aciego Pietri and Brokes (2008) mention that soil microbial biomass is optimum at pH level 7.00. The increase in soil pH to a neutral value tends to increase the content of soil microbial biomass which leads to higher content of organic matter. Thereafter, the greater amount of organic matter tends to increase total concentration of Cu in the soil. The relationship between total organic carbon and Cu in the soil can be seen in Hossam (2001) who found that increase in Cu associates with a corresponding increase in total organic carbon. According to Aciego Pietri and Brokes (2008) organic carbon reaches maximum value at around pH 5.00 and 6.00. This means the increase in soil pH to neutral value increases total organic carbon. Thereafter, the increase in organic carbon induces the increase of Cu content in the soil. Soil pH is related positively with Cu content in the form of logarithmic function and not in linear relationship due to other factors which can also affect the Cu content in the soil such as root growth, soil conditions, availability of N and presence of other nutrients (Wayland, 2010).

CONCLUSIONS

In acidic soil condition,soil microbial biomass C, biomass N, and Fe contents are related positively in the form of power function with soil pH and negatively with soil electrical conductivity. On the other hands, soil pH is related positively in the form of logarithmic function with the amount of Cu content in the soil. Soil pH does not form a well linear relationship with other selected soil properties due to the changes of these properties by the soil pH occurs indirectly in order to find the effect of soil pH on other soil properties in acidic soil at Ayer Hitam Forest Reserve.This study can be used as a part of soil fertility assessment especially in future tree planting program in Ayer Hitam Forest Reserve Puchong, Selangor, Malaysia.

REFERENCES

1. AciegoPietri, J.C. and Brokes, P.C., (2008), Relationships between soil pH and microbial properties in a UK arable soil. Soil Biology and Biochemistry 40, pp 1856-1861.
2. Bill. (2010), Iron (Fe++) [Electronic Version]. Spectrum Analytic Inc., 80, pp 321-1562.

3. Brady, N.C., Weil, R.R., (2002), The Nature and Properties of Soil, 13th ed. Springer Netherlands, p.249.
4. Bruckner M.Z., (2012), Water and Soil Characterization- pH and Electrical Conductivity, Microbial Life Educational Resources. Montana State University Bozeman, 2012.
5. Campbell C.A., Lafond G.P., Leyshon A.J., Zentner R.P. and Janzen H.H., (1999), Seasonal trends in soil biochemical attributes: effects of crop rotations in the semi-arid praire, Canadian Journal of Soil Science,79, pp 73-84.
6. Carroll, J.J. and Mather, A.A., (1992), The System Carbon Dioxide-Water and the Krichevsky-Kasarnovsky Equation. Journal of Solution Chemistry, vol. 21, pp. 607-621.
7. Drapper, N. R., and Smith, H., (1981), Applied Regression Analysis (2nd ed.), New York: John Wiley.
8. Fic, M. and Schroter, M.I., (1989), Batch studies for the investigation of the mobility of the heavy metals Cd, Cr, Cu and Zn. Journal of contaminated hydrology, 4, pp 69-78.
9. Flis, S.F., Glenn, A.R., Dilworth, M.J., (1993), the interaction between aluminium and root nodule bacteria. Soil Biology & Biochemistry, 25, pp 403–417.
10. Hanlon, E. A., Kidder, G. and McNeal, B.C., (1990), Soil-test interpretations and recommendation. Fla. Coop. Extn. Ser., IFAS, Univ. of Fla. Gainnesville, FL. Circular No. 817. p.49.
11. Helling, C.S., Chesters, C., Corey, R.B., (1964), Contribution of organic matter and clay to soil cation-exchange capacity as affected by the pH of the saturating solution. Soil Science Society of America Journal, 28, pp 517-520.
12. Hossam, M.A., (2001), Factors affecting mobility of Cu in soil-water matrices. Dissertation Sumitted to the Faculty of the Virginia Polytechnic Institute and State University in Partial Fulfillment of the Requirements of the Degree of Doctor of Philiosophy in Civil and Environmental Engineering. Blacksburg, Virginia.
13. Howell, J.M. and Gawthorne. J.M., (1987), Cu in Animals and Man.Volume II. CRC Press, Inc. Boca Raton, Florida.
14. Jelic M.Z, Milivojevic, J.Z., Trifunovic, S.R., Dalovic, I.G., Milošev, D.S. and Šeremešic, S.I., (2010), Distribution and forms of iron in the vertisols of Serbia. J. Serb. Chem. Soc. 76 (5), pp 781–794.

15. Juzefaciuk, G., Sokolowska, G., Sokolowski, S., Alekseev, A. and Alekseeva, T., (1995), Changes of minerological and surface properties of water dispersible clay after acid treatment of soils. Clay Minerals, 30, pp 149-155.
16. Kara, O. and Bolat, S., (2007), Impact of alkaline dust pollution on soil microbial biomass carbon, Turkish journal of agriculture & forestry,31, pp 181-187.
17. Kramme, A.D., (1985), Reference Manual for Assessing Water Quality Impacts from Highway Maintenance Practices. Washington, DC: US Department of Transportation, Federal Highway Administration.
18. Li, X.F. Li, B. Singh, Z. Rengelc, R. and Zhan, Z., (2007). Soil management changes organic carbon pools in alpine pastureland soils. Soil Till. Res, 93, pp 186–196.
19. Logah V., Safo, E.Y., Quansah, C. Danso, I., (2010), Soil Microbial Biomass C, N and P Dynamics Under Different Amendments and Cropping Systems in the Semi-deciduous Forest Zone of Ghana. West African Journal of Applied Ecology, 17.
20. Mitsios, I.K., Golia E.E., and Tsadilas C.D., (2005), Heavy Metal Concentrations In Soils And Irrigation Waters In Thessaly Region, Central Greece. Communications in Soil Science and Plant Analysis, 36, pp 487–501.
21. Okalebo, J.R., Gathua, K.W., and Woomer, P.L., (1993), Laboratory methods of soil and plants analysis: A working manual. Soil Sci. Soc. East African Tech. Pub. No.1 Marvel Epz, Nairobbi, Kenya. Community
22. Paiman B. Amat Ramsa, Y., (2007), Multimedia Super Corridor Heritage. Universiti Putra Malaysia Press.
23. Pathak, H., and Rao, D.L.N., (1998), Carbon and nitrogen mineralization from added organic matter in saline and alkali soils. Soil Biol. Biochem. 30, pp 695–702.
24. Potts, M.D, Kassim, A.R., Supardi, M.N.N., Tan, S. and Bossert, W.H., (2005), sampling tree diversity in Malaysian tropical forest: an evaluation of a pre-felling inventory. Forest Ecology and Management, 205, pp 385-395.

25. Provin, T. and Pitt, J.L., (2012), Managing Soil Salinity. Agricultural communication, The Texas A&M University System Web page: http://texaserc.tamu.edu Web page visited: March 2013.

26. Ristic, M., Bokic, T. and Zeceric, T., (2006). Cooper accumulation and availability in Vineyard soil of Serbia. Working and Living Environmental Reflection,3(1), pp35-42.

27. Rousk, R.,Baath E., Brookes, P.C., Lauber, C.L. Lozupone, C., Caporaso, J.G. Knight R.B. and Fierer N., (2010), Soil bacterial and fungal communities across a pHgradient in an arable soil. International Society for Microbial Ecology,10, pp 1751-7362.

28. Schulte, E.E., (2004), Soil and Applied Iron. Cooperative Extension Publisher. University of Wisconsin-Extension. United States

29. Shamshuddin, J. and Fauziah, C.1., (2010), Weathered tropical soils: the ultisols and oxisols. Malaysia: UPM Press.pp57-78.

30. U.S. Environmental Protection Agency (1997), Guidance Specifying Management Measurements for Sources of Non-Point Pollution in Coastal Waters, Washington, D.C.: Office of Water, U.S. Environmental Protection Agency.

31. United State Department of Agriculture (USDA), Soil Quality Indicator: Soil Electrical Conductivity; Natural Resources Conservation Services: United State, 2011.

32. Wardle, D.A., (1992), A comparative assessment of factors which influence microbial biomass carbon and nitrogen levels in soil. Biological Reviews of the Cambridge Philosophical Society 67, pp 321–358.

33. Wayland, (2010), Cu (Cu^{++}) [Electronic Version]. Spectrum Analytic Inc., 800: 321- 1562

Citations

CHAPTER 1

Mirza Hasanuzzaman, Kamrun Nahar, Md. Mahabub Alam, et al., "Potential Use of Halophytes to Remediate Saline Soils," BioMed Research International, vol. 2014, Article ID 589341, 12 pages, 2014. doi:10.1155/2014/589341.

CHAPTER 2

Ali, A., Perveen, S. , Muhammad Shah, S. , Zhang, Z. , Wahid, F. , Shah, M. , Bibi, S., and Majid, A. (2014) Effect of Foliar Application of Micronutrients on Fruit Quality of Peach. American Journal of Plant Sciences, 5, 1258-1264. doi: 10.4236/ajps.2014.59138.

CHAPTER 3

H.-L. Yu, W. Gu, J. Tao, J.-Y. Huang, And H.-S. Lin, "Impact Of Addition Of Fgdb As A Soil Amendment On Physical And Chemical Properties Of An Alkali Soil And Crop Yield Of Maize In Northern China Coastal Plain," Journal Of Chemistry, Article Id 540604, In Press.

CHAPTER 4

Yanhong Li, Huimei Wang, Wenjie Wang, Lei Yang, and Yuangang Zu, "Ectomycorrhizal Influence on Particle Size, Surface Structure, Mineral Crystallinity, Functional Groups, and Elemental Composition of Soil Colloids from Different Soil Origins," The Scientific World Journal, vol. 2013, Article ID 698752, 13 pages, 2013. doi:10.1155/2013/698752.

CHAPTER 5

Salihu Lukman, Mohammed Hussain Essa, Nuhu Dalhat Mu'azu, and Alaadin Bukhari, "Coupled Electrokinetics-Adsorption Technique for Simultaneous Removal of Heavy Metals and Organics from Saline-Sodic Soil," The Scientific World Journal, vol. 2013, Article ID 346910, 9 pages, 2013. doi:10.1155/2013/346910.

CHAPTER 6

Yong He, Lingling Hou, Hong Wang, Kelin Hu, and Brian McConkey, A Modelling Approach to Evaluate the Long-Term Effect of Soil Texture on Spring Wheat Productivity under a Rain-Fed Condition, doi:10.1038/srep05736.

CHAPTER 7

Walid Ellouze, Ahmad Esmaeili Taheri, Luke D. Bainard, et al, "Soil Fungal Resources in Annual Cropping Systems and Their Potential for Management," BioMed Research International, vol. 2014, Article ID 531824, 15 pages, 2014. doi:10.1155/2014/531824.

CHAPTER 8

Mohd-Aizat, A., Mohamad-Roslan, M.K., Wan Nor Azmin Sulaiman, and Daljit Singh Karam, The Relationship between Soil Ph and Selected Soil Properties in 48 Years Logged-Over Forest, doi: 10.6088/ijes.2014040600004.

Index

E

F

G

H

I

L

M

N

P

S